Neanderthals: Exploring our Genetic Past and Present

Neanderthals: Exploring our Genetic Past and Present

Written & Edited By:

Austin Mardon
Catherine Mardon
Lydia Sochan
Lilian Yeung
Shannon Lin
Sara Djeddi
Michaela Dowling
Ann Ping
Ariana Balassone
Romina Tabesh
Omar Abdul Hadi
Vedanshi Vala
Joylen Kingsley

Cover Design By:

Aleia Cabote

Copyright © 2021 by Austin Mardon

All rights reserved. This book or any portion thereof may not be reproduced or used in any manner whatsoever without the express written permission of the publisher except for the use of brief quotations in a book review or scholarly journal.

First Printing: 2021

Typeset and Cover Design by Aleia Cabote

ISBN 978-1-77369-635-5

Golden Meteorite Press
103 11919 82 St NW
Edmonton, AB T5B 2W3
www.goldenmeteoritepress.com

Table of Contents

Chapter 1:
The Study of Genetics and its Uses 09

Chapter 2:
The Origins and Impact of Discovering the Neanderthals 19

Chapter 3:
The Lifestyle of the Neanderthals 33

Chapter 4:
Neanderthals and The Development of Modern Cognition 43

Chapter 5:
Neanderthal Cranio-Facial Morphology 53

Chapter 6:
Ethical Principles of Genetic Research 65

Chapter 7:
Evolutionary Relationship Between Humans and Neanderthals 75

Chapter 8:
Genetic Links between Homo sapiens and Neanderthals 85

Chapter 9:
Ethnicity, Neanderthal DNA, and Medicine 95

Chapter 10:
The Neanderthals and Potential Causes of their Extinction 107

References 117

Chapter 1
The Study of Genetics and its Uses
by Lilian Yeung

Introduction

The study of genetics holds a long standing importance and interest in the scientific community, moreso in the medical field. With the understanding of genetics and human genomes, medical researchers are able to further improve quality of life for humans. The study of genetics also explains heredity by examining the DNA and components that are related to any and all aspects of genes (Durmaz et al., 2015). Before the study of the human genome, Gregor Mendel's experiments on cross-breeding pea plants was the breakthrough within science that widened the scope of human genetics, as he brought experimental genetics to a field dominated with mainly theoretical works (Durmaz et al., 2015). Mendel's work during the 1850s was revolutionary and continues to be a focal point in today's understanding of genetics. As the interest and need for experimental genetics increased, advancements in technology were also rising leading to more genetic studies conducted, furthering the understanding of genetics, particularly its molecular background. The Human Genome Project was a massive undertaking by researchers as unlike other simple organisms such as Drosophila (fruit flies) or E. coli, the human genome is complex and contains much more DNA to be analyzed. Researchers globally were competing with one another to be the first to completely sequence human DNA, which would be a massive achievement and provide the necessary starting point for many studies pertaining to understanding hereditary traits and diseases. But the study of genetics is not only relegated to the current human population; it can also be applied to fossils in order to determine their ancestry, health, and origins. Genetic studies can help shed light on the age of human fossils and answer questions as to how far removed they are from the current human population in terms of the human evolutionary tree.

Evolution of the Methods Used to Study Genetics

Within the field of genetics, there are many current techniques and methods that researchers use, but a crucial technological advancement was the invention of the single lens optical microscope in the late 1500s (Durmaz et al., 2015). Although there are many differing types of microscopes today, the invention of the single lens microscope has allowed geneticists to be able to visualize intracellular structures. This was a vital turning point in the field of genetics in that it allowed for an understanding of how cells worked along with the ability to visualize chromosomes.

As technological advancements were made in the field of science, experimental studies were also on the rise. One foundational experiment was Gregor Mendel's work with garden peas and their crossbreeding. It would be prudent to acknowledge that the pivotal work by Mendel was heavily influenced by Charles Darwin's work with regards to natural selection and breeding (Durmaz et al., 2015). Mendel's experiments consisted of crossbreeding pea plants and exploring how traits were inherited and expressed. His work in 1865 was fundamental in creating the basis for the law of inheritance (Durmaz et al., 2015). It led to improvements on future genetic studies, such as Thomas Hunt Morgan's experiments in 1910 which confirmed that the appearance of a certain phenotype in a chromosome was due to genes (Durmaz et al., 2015).

Following shortly after that experiment was the sequencing of genes of organisms. The first organism to have had its entire genome sequenced and thus, created a genetic map, was Drosophila (fruit flies) (Durmaz et al., 2015). Though many genetic studies have been made, there was still much to be discovered as the genetic structure of hereditary material was still unknown (Durmaz et al., 2015). It was important to understand the genetic structure of hereditary material as it is connected to the inheritance of traits throughout generations. Thus, a major endeavor was undertaken to sequence the entire human genome, a key component to understanding DNA. Watson and Crick's work studying the human genome led to the realization that DNA was helical and double-stranded with complementary strands that were antiparallel to one another (Durmaz et al., 2015). With this discovery, much interest was placed on decoding human DNA to understand its meaning. Not only would the results further our understanding of genetics and human DNA, it would also shed light on understanding human evolution, diseases and their genetic origins, in addition to providing information to the 'nurture vs. nature' debate that has been occurring for centuries (Durmaz et al., 2015). This resulted in the Human Genome Project, a massive undertaking by numerous researchers with the goal to fully sequence the entirety of the human DNA by determining its nucleotide sequence (Venter et al., 2001).

Human Genome Project

The Human Genome Project consisted of the entire human DNA being decoded, which has many benefits to the scientific community as a whole. It would help clarify our current understanding of the human DNA along with hereditary genes and traits. This project started after the established technique of DNA sequencing, a method developed by Sanger in 1977 where the order of DNA nucleotides was determined with the use of nucleotide analogs—nucleotides with an extra sugar and phosphate attached to it (Venter et al., 2001). Researchers first began preparing for the project by collecting samples from 5 different individuals from different ethnic backgrounds (Venter et al., 2001). This allowed them to sequence a complete human genome which should be composed of multiple ethnic backgrounds for diversity along with building a higher quality of DNA libraries (Venter et al., 2001). Once the collection of samples were taken from these volunteers, researchers began to do a shotgun sequencing on their entire genome (Venter et al., 2001). This method includes the breakage of the DNA genome into randomized tiny fragments which are then individually sequenced. With shotgun sequencing, much time is saved on sequencing as it allows simultaneous sequencing of multiple tiny fragments, but the tradeoff to this method is the lack of precision. In terms of precision, shotgun sequencing would not be able to assemble the sequenced fragments in a complete fashion and would require the manual assistance of researchers to oversee this process even with the use of computer algorithms (Venter et al., 2001). In addition, shotgun sequencing may have trouble assembling repeated sequences and also needs a reference genome when processing (Venter et al., 2001). Furthermore, due to the speed of the sequencing, it may also introduce more errors during the assembly processing (Venter et al., 2001). Because of the importance of sequencing the human genome, and the delicate interplay between efficiency and accuracy, there were numerous methods used to assemble the human genome (Venter et al., 2001). A system developed to help identify and characterize human genes was called Otto (Venter et al., 2001). This rule-based system is a software that is essentially an annotator and helps to identify a region of the genome, in which a human curator would then examine and analyse how different regions of a genome were related to one another (Venter et al., 2001). Overall, once the human genome mapping was completed, it paved the way towards many other studies into genetic diseases and traits, their heritability, and how the genome mapping could help improve the quality of life for humans along with our understanding of human genes.

Tools and Methods Used to Study Genetics

The study of genetics encompasses many tools and methods that are widely used within the field of science, such as the microscope, Northern, Western, and Southern blotting techniques in addition to polymerase chain reaction techniques. In the field of cytogenetics, which is the field of genetics looking at inheritance in relation to the function and structure of chromosomes, staining methods and genetic analysis were conducted leading to the discovery of the total number of human chromosomes, which is 46 (Durmaz et al., 2015). When studying genetics, there are many ways in which this can be done. The sample that is taken from an organism can be from any cell or body parts. In the case of humans, samples can be taken from bodily fluids (blood, saliva, semen, etc.), hair, finger nails, skin cells, along with many other parts of the body (Venter et al., 2001). Such a variety of samples is useful in the ability to correctly identify age, ethnicity, ancestry, along with potential hereditary diseases including cancer and dormant diseases (Durmaz et al., 2015).

A useful method of studying genetics is the identification of human chromosomal abnormalities within pregnancies. This is done through culture methods of peripheral leukocytes that also includes staining and fixation methods (Durmaz et al., 2015). Once this culture method was established, it allowed for the discovery of abnormal human chromosomes linked to certain congenital defects such as Down syndrome, Klinefelter and Turner syndrome, along with multiple sex chromosome abnormalities where there may be missing or additional parts or whole sex chromosomes (Durmaz et al., 2015).

In the past, genetic techniques for the analysis of fetus samples could only be achieved postnatally as there were not any safe alternatives to collecting fetal samples during human pregnancy (Durmaz et al., 2015). But after 1966, prenatal samples were then able to be collected through the amniotic fluid which would provide fetal cells (Durmaz et al., 2015). This method, called amniocentesis, is an invasive procedure where a needle is inserted into the amniotic fluid of the uterus to collect a sample in which developing abnormalities would be screened for. Amniocentesis is quite important in providing information on a developing fetus and whether any alterations or concerns should be addressed in terms of the health of the mother and child.

In 1976, a novel banding technique was established by Yunis which improved resolution for banding techniques that used lymphocyte cultures (Durmaz et al., 2015). Banding techniques are genetic analyses that examine the bandings found on chromosomes and use diagrams to show the different characteristics of the bands on each chromosome (Durmaz et al., 2015). Furthermore, the characteristics of each chromosomal band may serve as an indication of latent diseases or susceptibility for certain ones (Durmaz et al., 2015). With any banding techniques, the resolution of diagrams is entirely dependent upon the type of banding method used where methods used primarily in the early to late 1900s were of lower resolution (Durmaz et al., 2015). But with technological advancement, higher resolution techniques are being discovered, helping to identify more chromosomal abnormalities and visualize chromosomes. Yunis' banding technique was vital in the identification of genetic abnormalities including Wolf-Hirschhorn and Cri-du-Chat syndromes which is when parts of a chromosome are missing (Durmaz et al., 2015).

A vital procedure developed to examine cytogenetics and molecular genetics was called fluorescence in situ hybridization (FISH). FISH method identifies particular sequences of nucleic acids, useful in identifying certain genes or chromosomes (Durmaz et al., 2015). Although many years have passed and there have been many significant improvements made to this technique, the FISH method has laid the foundation for current technology analysing specific DNA sequences and labelling them with fluorescent probes (Durmaz et al., 2015). Present-day technology continues to use fluorescence in many of their probing methods, such as in gold-based probe or enzyme-based probe systems (Durmaz et al., 2015). In addition, labelling of sections of chromosomes was enabled with the technology such as nick translation, a process of radioactively labelling DNA strands (Durmaz et al., 2015). Nick translation and newer technology uses probes, whether they be fluorescent probes or perhaps some other radioactive probe, to bind to DNA bases which are used as a marker for that specific gene.

Conclusion

This chapter explores a brief history of the study of genetics and some of the major contributions and technology that were part of the genetics field. A notable experiment that facilitated the beginning of experimental genetic studies was Mendel's work covering pea plants and the effects of cross breeding them. With the advancement of technology coinciding with the rise in genetic studies, it resulted in many novel experiments, technologies and systems being created and refined over time. Such experiments included the FISH method along with banding techniques. With each establishment and refining of genetic study techniques, more and more chromosomal abnormalities are being discovered and therefore allow doctors to better be able to treat patients affected from such malformations. It is important to note that genetic studies and techniques should always continue to improve, and pave the road for new methods to be established that would allow a better understanding of the origins of chromosomal abnormalities as mutations are always arising within the human genome. As researchers have a better understanding of chromosomal abnormalities and where they are located in the genome, more research needs to be made in treating these malformations and preventing them from occurring in the first place.

Chapter 2
Universal Traits of Domestication
by Shannon Lin

Introduction

Since its discovery in a German cave, findings of Neanderthal fossils have incentivised both past and present scholars to explore their origins and their impact on human evolution (Madison, 2016; Tattersall & Schwartz, 1999). Known to have relational DNA threaded within modern day humans, the Neanderthals have brought forth a notable set of anatomical, dietary, and behavioural characteristics that show them to be more advanced than what was once thought to be (Gee, 2000; Hublin, 2009; Madison, 2016; Mellars, 1998; Soressi, 2016; Tattersall & Schwartz, 1999; Weyrich et al., 2017). To be discussed further throughout this chapter, the complexity of this evolutionary cohort has been noteworthy to understand how their classification as an 'other' is rather a unique representation of various skill sets that are necessary for survival (Madison, 2016; Mellars, 1998). By studying the history of the Neanderthals and the impact they have made on their successors, this allows for further inquiry and diversity in anthropological research that can be highly beneficial in all aspects.

The Impact of its Discovery and Origins

Originally discovered in a German limestone quarry in August 1856, a fossilized skullcap and partial skeleton unveiled from the Feldhofer Cave quickly gained vast attention from scientists and scholars alike, and this was acknowledged as the "first scientifically recognized Neanderthal" (Madison, 2016, p. 411; Tattersall & Schwartz, 1999). Its lineage has been prescribed to inhabit Europe and Western Asia between approximately 200 and less than 30 thousand years ago while the Neanderthal population was replaced by modern populations about 30,000 to 40,000 years ago (Madison, 2016; Mellars, 1998; Soressi, 2016). Suggested by Hublin (2009), this was likely due to the recolonization of Western Eurasia by hominins of African origin into Europe, where the author stressed the significance of the close phylogenetic relationship between the Neanderthals and Homo sapiens. This argument is further supported by the fact that the Neanderthals coexisted and interbred with anatomically modern humans across Eurasia during the Late Pleistocene epoch, therefore establishing them as our closest-known hominin relatives (Hublin, 2009; Weyrich et al., 2017). Currently known as our first extinct human relative with substantial fossil and behavioural records, the discovery of the Feldhofer skeleton from the Neander Valley has truly called for extensive research and studies that examine the origins and lifestyle of the Neanderthals from various perspectives (Hublin, 2009; Madison, 2016; Tattersall & Schwartz, 1999).

Upon their discovery, the Feldhofer fossils sparked an array of curiosity from early scientists between 1856 and 1864, where several findings were made (Madison, 2016). First to examine the fossils, Johann Carl Fuhlrott the naturalist described the forehead to be low and flat, which contrasts with the high doming that occurs in those of most humans (Madison, 2016). Following this, Hermann Schaaffhausen found the skull to take on a long and elliptical form, while having an 'unusual thickness' for a human, and most notably identified its prominent brow ridge (Madison, 2016). This brow ridge was 'enormously great' and it was an indication of the result of an expanded frontal sinus to enhance oxygen intake for increased movement force and power of endurance (Madison, 2016). Schaaffhausen added that this thickness and rigidity was also found in its 'unusually rounded' ribs and thick limb bones that supported a large amount of muscle mass (Madison, 2016). Assisted by anatomists George Busk and Thomas Henry Huxley, the trio formed by British geologist Charles Lyell reiterated the striking presence of the brow ridge, which led to the conclusion that such a feature was highly comparable to the skulls of apes (Madison, 2016). This was also agreed upon by Schaaffhausen, where Charles Carter Blake of the London Institution furthered this comparison to resemble the skull of the gorilla (Madison, 2016). Geologist William King of Queen's College Galway in Ireland described the Neanderthal skull as more ape-like rather than man-like, and suggested a distinct classification within the human genus to be named Homo neanderthalensis (Madison, 2016).

What Makes the Neanderthals Different

While the previous findings have indeed established the Neanderthal to have distinct features such as the prominent brow ridges, flat and oblong cranium, ape-like characteristics, and a short, stocky, and robust morphology, these features have ultimately drawn attention toward the Neanderthals to be identified as an 'other' during paleoanthropology's history (Madison, 2016; Mellars, 1998). Despite their brain size that fell within the human range, this classification of the 'other' creates the implication that the Neanderthals lack various human aspects, which establishes their status as lesser than that of the Homo sapiens (Madison, 2016). This is exemplified through the differences between the Neanderthals and modern humans in terms of their brain shape, developmental patterns, obstetric features, and potentially their life history (Hublin, 2009). As this questions the social and behavioural aspects of the Neanderthals, Mellars (1998) proposed that their morphology has rather made functional contributions that may have been necessary for survival. Specifically, the author suggested that the pressures of heavy chewing and usage of the jaw as a tool led toward their iconic depiction of beings with large noses and faces. Furthermore, their massive body weight and musculature in the upper limb bones was a reflection of exceptionally heavy stresses on the arms and hands (Mellars, 1998). Considering how the Neanderthal morphology was influenced by a high degree of physically taxing labour, it is then important to assess how their diet accommodated this lifestyle and upheld their survival.

Dietary Components of the Neanderthals

Emphasized by Hublin (2009), the Neanderthals were large-bodied and required greater amounts of energetic intake compared to recent hunter gatherers. This assumption was supplemented by isotopic studies that found their primary protein source to come from animals (Hublin, 2009). Additionally, archaeological records of Neanderthals found them to be efficient hunters that targeted medium to large mammals for meat and fat (Hublin, 2009). Archaeological and isotopic data from the last glacial cycle in particular have suggested that the carnivorous diet of the Neanderthals were comparable to polar bears or wolves (Weyrich et al., 2017). Primarily, their diet seemed to consist mainly of large terrestrial herbivores including reindeer, woolly mammoth, and woolly rhinoceros (Weyrich et al., 2017). While this was shown in the oral bacterial community (microbiota) samples from the Spy Cave in Belgium, these variations in microbiota and DNA from calcified dental plaque have established regional differences in Neanderthal ecology that must be investigated further to gain a well-rounded understanding of the Neanderthal diet (Weyrich et al., 2017).

Calculus, which refers to calcified dental plaque, provides a wealth of information regarding the behaviour, diet, and health of ancient hominin species (Weyrich et al., 2017). Its preservation system can be used to analyze different microbial species and how they have evolved and spread among hominins (Weyrich et al., 2017). These dietary differences were found to directly relate with the overall shifts in microbiota, indicating that the consumption of meat contributed to a large degree of variations across Neanderthal microbiota (Weyrich et al., 2017). According to Weyrich and colleagues (2017), these differences are highly influenced by the ecological settings that the Neanderthals had once inhabited. As such, the Neanderthal diets and food choices were guided by what was local and most readily available (Weyrich et al., 2017).

Found in the Spy Cave samples in Belgium, the Spy II Neanderthal diet was heavily reliant on meat (Weyrich et al., 2017). This was presented through the evidence of woolly rhinoceros, reindeer, mammoth, and horse bones in Spy Cave, which confirms the isotope and dental microwear data from the Spy individuals that inferred a carnivorous lifestyle (Weyrich et al., 2017). Weyrich et al. (2017) also found calculus evidence to be closely linked with the regular consumption of mushrooms, specifically of the grey shag variety. From the samples found in the El Sidrón cave in Spain, mushrooms were also a primary dietary component, along with pine nuts and moss that reflected forest gathering (Weyrich et al., 2017). Contrasted by the Spy Cave populations, Weyrich and colleagues (2017), however, did not find evidence that tied the El Sidrón Neanderthals to large herbivores or high meat consumption. Other indicators of herbivorous lifestyles amongst the Neanderthals are shown through phytoliths, starch granules, and proteins that are prevalent in calcified dental calculus (Weyrich et al., 2017). It is also suggested that the variety of plants that constitute the Neanderthal diet included plants that are rather used for medicinal purposes (Weyrich et al., 2017). One example would be the findings of calculus sequences that corresponded to poplar, which was associated only with the El Sidrón individual (Weyrich et al., 2017). Poplar contains salicylic acid, a natural painkiller, which is also an active ingredient in aspirin (Weyrich et al., 2017). The evidence from the samples also contained sequences of the natural antibiotic producing Penicillium, which was obtained from the moulded herbaceous material (Weyrich et al., 2017). These discrepancies in diet shows how the Neanderthals had truly relied on what was most ecologically available to them. This is also made clear in the differing lifestyles and survival mechanisms across different regions of Neanderthal inhabitation.

Lifestyles and Survival Mechanisms

The varying dietary components and lifestyles of the Neanderthals has demonstrated their presence across borders, which draws further inquiry into their demographics and differing lifestyles. As such, Neanderthal research has become elevated through the notable findings of early underground constructs (Soressi, 2016). Precisely, the findings of 175,000 year old structures inside a cave in southwest France were attributed to the Neanderthals since they were the only hominin group that was present in western Europe during that time (Soressi, 2016). Suggesting that the Neanderthals have explored the underground and advanced their construction abilities, these findings therefore establish a more complex social behaviour than what was previously thought of them (Soressi, 2016).

Citing the findings by Jaubert and colleagues, Soressi (2016) described these structures to measure up to 40 centimetres high and 6.7 metres wide. These circular structures, which were made up of accumulations of broken stalagmites, were shown to be heated by small fires due to their reddened, blackened, and cracked characteristics (Soressi, 2016). While two of the structures were semicircular in shape, these collective accumulations of almost 400 stalagmites and its fragments constructed walls that consisted of 4 superimposed layers of stalagmite fragments, approximately 30 centimetres in length (Soressi, 2016). Smaller elements were encapsulated obliquely in between each layer while the structures were quickly sealed by calcite after their completion (Soressi, 2016). As a result, these structures are considered to be the best-preserved constructions out of the entire Pleistocene epoch (Soressi, 2016). These structural findings strongly demonstrate how broad-scale and time-specific reconstructions of ancient history can be made possible (Soressi, 2016). Through the findings of skeletal remains and fossils, this can provide further insight into the overall lifestyle of the Neanderthals while studying these differences across various regions.

Discussed in Mellars (1998), a number of archaeological findings have brought forth explanations that pertain to Neanderthal behaviour. Using radiocarbon and uranium series data, these date a Neanderthal lower jaw that was obtained from Andalusia, which is located in the southern part of the Spanish peninsula, south of Ebro valley (Mellars, 1998). From the site of the Zafarraya in Andalusia, this sample was found to have existed around 30,000 years before present, and the Neanderthals were likely to have survived for a minimum of 5,000 to 10,000 years after the arrival of modern populations (Mellars, 1998). On the contrary, the Neanderthal population from the region north of the Ebro valley adapted to behaviours more alike the first anatomically modern populations despite their existence around 38,000 to 40,000 years before present (Mellars, 1998). In the north, this group in particular was described to survive upon the production of simple bone tools, perforation of animal teeth along with other forms of personal decoration, as well as the scattering of powdered red ochre across the floors within their living areas (d'Errico et al., 1998 & Bahn, 1998, as cited in Mellars, 1998). Considering these differing behaviours and developments, Mellars claimed that the ecological differences across the northern and southern regions of the Iberian Peninsula are the most likely explanation for their extensive period of coexistence.

On the other hand, there are high densities of archaeological sites that have been placed in western France (Mellars, 1998). Exceptionally high in population, when coupled with the cooler conditions opposite the southern Mediterranean sites, this combination likely resulted in the delay in colonization by modern human groups (Mellars, 1998). With cooler climates and open vegetation, these conditions may have seemed to contribute toward the relatively short and stocky bodies that Neanderthals are known for (Mellars, 1998). From this perspective, it would mean that the Neanderthals underwent climatic adaptation to survive in such temperatures (Mellars, 1998). This would be showcased through the need to warm up cold air streams into the nasal passages, which resulted in the large noses and faces that the Neanderthals are known for (Mellars, 1998). Mellars (1998) also argued that this facial structure was the result of the pressures of heavy chewing and the need to utilize their jaws as tools. Instead, these functional pressures were rather used to perform several survival mechanisms, which would make the most sense in environments that ranged in a variety of climates such as the Eurasia region (Hublin, 2009; Mellars, 1998). While the majority of their ecological range remained within the relatively temperate areas of southern or western Eurasia, and limited on peri-arctic landscapes, Hublin (2009) emphasized that the Neanderthal body proportions are determined by the technical limitations when responding to cold climatic stresses (Mellars, 1998). Through such evidence and studies, it is understood that their distribution throughout different regions took upon variations in ways of living. As suggested by Hublin, "these changes in geographical distribution primarily resulted from local extinctions rather than from migration and habitat tracking" (p. 16022). By understanding this, archaeological, paleontological, and paleogenetic evidence must be looked at to assess how the Neanderthals left their footprints and marked new ones as time progressed (Hublin, 2009).

Neanderthal DNA in Other Regions and their Disappearance

The distribution of the Neanderthal population has indeed questioned how their DNA has interacted with the introduction of anatomically modern humans. By undergoing long-term hybridization, Tattersall and Schwartz (1999) argue that doing so would undoubtedly categorize the two populations to be within the same species as a distinct, yet odd, form of Homo sapiens. This turning point occurred due to the arrival of the first modern humans in Europe, the Homo sapiens, which disrupted the Neanderthal population after a substantial period of inhabitation (Tattersall & Schwartz, 1999). This is demonstrated through the skeletal findings of a 4-year-old child that was unearthed at a 24,500 year old site in Portugal (Tattersall & Schwartz, 1999). This evidence from Duarte et al.'s findings in Tattersall and Schwartz' commentary strongly suggests that this "represents not merely a casual result of a Neanderthal or modern human mating, but rather is the product of several millennia of hybridization among members of the resident Neanderthal population and the invading Homo sapiens" (Tattersall & Schwartz, 1999, p. 7117).

Several explanations for this generational impact can therefore be concluded from this statement alone. Firstly, the Neanderthals could have been eliminated through direct conflict or indirect economic competition with the moderns (Tattersall & Schwartz, 1999). This explanation implies the separate species status of the Neanderthals, while indicating species continuity (Tattersall & Schwartz, 1999). Another possible reason for their eviction may come from the idea that the Neanderthals rapidly evolved into modern humans, or the genes of the invading modern humans took over those of the Neanderthals (Tattersall & Schwartz, 1999). Again, this indicates some form of species continuity, which puts into question whether there is Neanderthal DNA transcribed into today's DNA in modern humans (Tattersall & Schwartz, 1999). According to Gee (2000), the geographical age of the Neanderthal from Feldhofer is undetermined. However, the finding of a fossil from a region 2,500 kilometers from Feldhofer, named the Caucasus Neanderthal, has suggested that these two individuals have lived tens of thousands of years apart (Gee, 2000). When compared with the DNA of modern humans, the DNA from both of these samples have been established as distinct and unique from humans regardless of their racial or geographic origin (Gee, 2000, para. 6). It can be concluded that within modern humans, many would carry this

DNA derived from these archaic populations that interbred during the Late Pleistocene (Rogers et al., 2017). To be discussed further in Chapters 7 and 9, it is of utmost importance to recognize how the Neanderthal genetic composition has shifted over time to gain a comprehensive understanding of human evolution.

Conclusion

Altogether, these findings have undoubtedly shed light on the fascinating history and origins of the Neanderthals. Through a combination of research on the Neanderthal morphology, diet, lifestyle, demographics, and contact with anatomically modern humans, these studies have nonetheless provided a fruitful evaluation of how they lived and survived across several regions and time periods (Gee, 2000; Hublin, 2009; Madison, 2016; Mellars, 1998; Soressi, 2016; Tattersall & Schwartz, 1999; Weyrich et al., 2017). Neanderthals have been viewed as an 'other' species, due to their ape-like cranial characteristics and their short and stocky figures (Madison, 2016; Mellars, 1998). Through isotopic and calculus studies, it has been discovered that the Neanderthals mainly followed a carnivorous diet rich in reindeer, rhinoceros, and woolly mammoth, with the exception of El Sidrón Neanderthals who consumed a plant-based diet heavy in mushrooms and pine nuts (Hublin, 2009; Weyrich et al., 2017). Underground findings of constructed stalagmite structures indicate that the Neanderthals attained a versatile behavioural repertoire (Mellars, 1998; Soressi, 2016). By assessing the similarities and differences between the Neanderthals and the modern human population, such research can evaluate the state of human evolution for man that walked this Earth over 30,000 years ago (Madison, 2016; Mellars, 1998; Soressi, 2016).

Chapter 3
The Lifestyle of the Neanderthals
by Sara Djeddi

The human species Homo neanderthalensis, more commonly known as Neanderthals, lived approximately 400,000 years ago until they abruptly disappeared 40,000 years in the past (Mellars, 2004). These hominids are characterized by their larger skulls, bigger noses, shorter height, and stockier bodies in relation to Homo sapiens (Appenzeller, 2013). Moreover, Neanderthals lived mainly in Europe and Southwest Asia, which were later inhabited by modern humans 50,000 years ago (Mellars, 2004). Throughout the years, using fossil remains as well as genetic analysis, researchers were able to establish details about their lifestyle, behaviour, and how they managed to survive.

Tools

Neanderthals were alive during a period of time consisting of exceptionally harsh environmental conditions (Appenzeller, 2013). For the most part, they lived throughout Europe during the unpleasant circumstance of the Ice Age, which was key in leading to human adaptations (Appenzeller, 2013).. In order to ensure survival, they developed intricate tools for hunting and domestic purposes (Douka & Spinapolice, 2012).

A large quantity of the tools discovered from the Palaeolithic sites were made of stone (Douka & Spinapolice, 2012). In addition to this, researchers found shell tools which they believe were made due to a lack of raw material (Douka & Spinapolice, 2012). The Mousterian industry is the culture of the Neaderthal stone tools, specifically in Europe, Western Asia, and Northern Africa approximately 40,000 years ago (Douka & Spinapolice, 2012). The manner in which their tools were assembled involved flaking techniques, which is done by adding pressure and force to the edge of such tools as the final touches (Douka & Spinapolice, 2012). The flaking was accomplished some of the time with hammer stones, animal bones, or teeth (Douka & Spinapolice, 2012). Examples of the stone tools include hand axes and triangular side scrapers (Douka & Spinapolice, 2012). Other fossils and remains imply that Neanderthals used their front teeth as tools, essentially as an extension of their hands (Douka & Spinapolice, 2012).

Along with Neanderthal remains and fossils, artifacts that are non-functional were discovered, indicating that their cognition surpassed that which is needed for basic survival (Rodríguez-Hidalgo et al., 2019). On this basis, it was suggested that they collected artifacts for other purposes, such as beauty (Rodríguez-Hidalgo et al., 2019). Additionally, they made tools that were used domestically, such as those for food preparation, sharpening, cutting through bone and wood, hole punching, and scrapping tanning hides (Rodríguez-Hidalgo et al., 2019).

Hunting

The hunting strategies and tactics developed by Neanderthals were studied using a sequence found in their shelter, known as Abric Romaní (Marín, 2017). This sequence is 15 000 years old, which is incredibly useful in the determination of how Neanderthals hunted while occupying this shelter (Marín, 2017). Using the principles of ecologists, the researchers determined animals' ages of death which was useful in population dynamics via life tables (Marín, 2017). Specifically, this research looked at bison kill sites to determine the time frame of predation and hunting (Marín, 2017). In order to investigate the animals' age at death, researchers focused on the arrangement of their teeth (Marín, 2017). This was done with the use of two methods, non-destructive and destructive (Marín, 2017). Both of these are essentially comparing the state of the animals' teeth with those of animals with ages at death that are known (Marín, 2017). The non-destructive method includes taking measurements of the teeth or analyzing the surface wear (Marín, 2017). But, an example of destructive analysis is cementochronology, where layers of cement deposits are counted (Marín, 2017).

Based on the evidence gathered from the Abric Romaní sequence, it was established that the Neanderthals used selective and non-selective techniques for hunting (Marín, 2017). The selective strategy focused on catching older prime animals, which was possibly accomplished using ambushes in order to choose specific prey (Marín, 2017). In contrast, non-selective hunting involved consuming animals of any age focusing more on abundance, which was done using encounter hunting or ambushes that are non-selective (Marín, 2017). Additionally, other more specific tactics were likely used that were specified based on the type and age of the prey that would be pursued (Marín, 2017). For instance, in order to catch a large animal, strategies involving group cooperation would have been used (Marín, 2017). Due to the variability, it is not useful to associate Neanderthals of Abric Romaní with a single technique or type of hunting (Marín, 2017).

Genetics

Research involving the genetics of Neanderthals and Homo sapiens has greatly contributed to understanding of Neanderthal extinction about 35,000 years ago (Mellars, 2004). Additionally, studies of DNA have determined that the genetic origin of modern humans is in Africa 150,000 years ago, and that their migration to other areas began 30,000 to 60,000 years ago (Mellars, 2004). Moreover, Neanderthals and modern humans are thought to have first come in contact with each other 55,000 in the past; this date is based on the discovery of a human skull in Galilee, Israel (Mellars, 2004).

A great deal of the genetic research in Europe is based on the mitochondrial DNA of Neanderthal skeletal remains and early H. sapiens (Mellars, 2004). Interestingly, it was found that the mitochondrial DNA of Neanderthals that lived in Europe differs vastly from both current H. sapiens and early modern humans that lived in the same region (Mellars, 2004). This variance strengthens the idea that there was a very small amount of interbreeding in Europe between Neanderthals and modern humans, hence the lack of genetic trace in the European gene pool (Mellars, 2004). Contrary to the aforementioned fossil evidence, the Neanderthal mitochondrial DNA analyzed shows that the ancestors of the modern human and the Neanderthals split from one another approximately 300,000 years ago (Mellars, 2004).

Neanderthals and modern humans also share genes, and some of these genes have implications related to the immune system (Yotova et al., 2011). Specifically, a part of the modern human dystrophin gene in the X chromosome, called dys44, is also present in Neanderthals (Yotova et al., 2011). Dys44 has a group of alleles that are inherited together, known as B006 (Yotova et al., 2011). This region was found to be related to the interbreeding of the two hominids back 80,000 to 50,000 years ago (Yotova et al., 2011). Moreover, some additional genes that were inherited from Neanderthals could be involved in the skin tone of those living in East Asia, or even play a role in the risk of developing type 2 diabetes (Yotova et al., 2011).

The comparison of Neanderthal and modern human genetics can also be useful in determining past behaviour. This type of research has become increasingly prevalent since 2010 due to the sequencing of the complete genetic sequence of Neanderthals. Through genetic evidence, physical characteristics of this species as well as their lifestyle were able to be uncovered. Some Neanderthals were thought to have red hair and pale skin, and live in small family groups that are isolated (Ledford, 2007).

Diet

In a 2017 study, DNA from Neanderthal dental calculus, also known as calcified dental plaque, was sequenced in hopes of learning about behavior (Weyrich et al., 2017). This type of genetic analysis can be very useful in examining the health implications of the possible interbreeding of Neanderthals and modern humans in Eurasia (Weyrich et al., 2017). The dental calculus preserved in the Neanderthals' DNA is useful in providing insight on their health, behaviour, as well as microbiota evolution and dissemination (Weyrich et al., 2017). Based on past archaeological data, it is thought that Neanderthals had a large intake of meat as part of their diet, specifically from large herbivores such as woolly rhinoceros (Weyrich et al., 2017). However, topics surrounding Neanderthal diet and health are still subject to debate due to the limited information regarding their dietary intake and its relation to disease (Weyrich et al., 2017).

The first analysis involved dental calculus from Neanderthals that lived in caves in Spain, Belgium, and Italy (Weyrich et al., 2017). The individuals' remains were compared to both modern humans and chimpanzees with the intention of determining Neanderthals' diet (Weyrich et al., 2017). The calculus of the Neanderthals in the Belgian cave showed they had a diet consisting highly of meat, specifically woolly rhinoceros (Ceratotherium simum) and wild sheep (Ovis aries) (Weyrich et al., 2017). Additionally, it showed that edible grey shag mushrooms (Coprinopsis cinerea) were a large part of their diet (Weyrich et al., 2017). Both of these findings supported the previous isotopic and dental evidence that Neanderthals in these regions had a carnivorous diet with the exception of their consistent mushroom intake (Weyrich et al., 2017).

On the other hand, the dental calculus from Neanderthals that lived in El Sidrón, Spain showed a different dietary composition. They included traces of edible mushrooms (split gill; Schizophyllum commune), pine nuts (Pinus koraiensis), forest moss (Physcomitrella patens), and poplar (Populus trichocarpa) (Weyrich et al., 2017). In addition to those, the dental calculus showed plant fungal pathogens including Zymoseptoria tritici, Phaeosphaeria nodorum, Penicillium rubens, and Myceliophthora thermophila, indicated that there was possibly some intake of moulded plants (Weyrich et al., 2017). Moreover, other analyses of tooth remains from different regions demonstrate that Neanderthal diet is based upon what food was locally available (Weyrich et al., 2017).

Interestingly, dietary intake has an impact on the microorganisms found within hominins, meaning that the latter found in dental calculus could be used to obtain information on the diet and behaviour of human species (Weyrich et al., 2017). In the case of Neanderthals, it was found that their microbiota is more comparable to historic chimpanzees rather than to modern humans (Weyrich et al., 2017). Specifically, their microbiota was composed of less Gram-negative pathogens, which can be involved in periodontal disease (Weyrich et al., 2017). In other words, Neanderthals' microorganisms contained less bacteria that cause infection of teeth and gums.

Proteins and starches of the Neanderthal dental calculus were also investigated, showing that plants were consumed for dietary and medicinal reasons (Weyrich et al., 2017). For instance, it was suggested that a Neanderthal from El Sidrón may have self-medicated because of a dental abscess (Weyrich et al., 2017). This was inferred due to the presence of salicylic acid sequences in the individual's calculus, which is a pain-killer present in aspirin (Weyrich et al., 2017). There were also sequences of antibiotics as well as a gastrointestinal pathogen (Enterocytozoon bieneusi) causing diarrhoea, which may be another cause of their self-medication (Weyrich et al., 2017).

Culture and Art

The oldest painting, a red dot, was found in El Castillo cave about 41,000 years ago (Appenzeller, 2013). This timeline aligns with the arrival of modern humans in western Europe, however it is argued that the painting is much older and was created by Neanderthals (Appenzeller, 2013). This prompts the following question: "did the Neanderthals, once caricatured as brute cavemen, have minds like our own, capable of abstract thinking, symbolism and even art?" (Appenzeller, 2013, p. 1)

In the south of Spain, a few cockle shells were discovered (Appenzeller, 2013). Interestingly, these shells were covered in holes along their edges, almost as if they were used as ornaments, or possibly even worn by Neanderthals (Appenzeller, 2013). A few of the uncovered shells had colourful pigments on them, as though they were containers (Appenzeller, 2013). These shells are dated as approximately 50,000 years old, which is prior to the arrival of modern humans in the area (Appenzeller, 2013). This could be an example of symbolism and deeper thinking amongst Neanderthals.

It has been suggested that Neanderthals were also somewhat spiritual, due to the evidence alluding to their burial of the dead (Appenzeller, 2013). In addition, it was found that they made glue for their weapons in a unique manner that is difficult to duplicate in the present day (Appenzeller, 2013). Other sites had pigments in them, almost like crayons (Appenzeller, 2013). Using pigments such as red ochre and black manganese, it is thought that Neanderthals painted symbolic patterns on themselves, which could also be signs of spirituality and symbolism (Appenzeller, 2013).

The complete Neanderthal genome released in 2010 demonstrated that it differed from modern humans in areas related to brain function, specifically in regions used for vision and body control (Appenzeller, 2013). Neanderthals had more robust builds, they were more broad, which required different demands from their brains as compared to modern humans (Appenzeller, 2013). This could have also resulted in the latter being less socially aware and causing them to interact differently than modern humans (Appenzeller, 2013).

Although there are still uncertainties regarding Neanderthals and their way of life, there have been numerous discoveries of their remains, leaving room for further research and investigation. The fascination with their lifestyle persists, as they are the closest relatives to modern humans (Appenzeller, 2013).

Chapter 4
Neanderthals and The Development of Modern Cognition
by Michaela Dowling

The Study of Cognition

The study of modern behavior's origin amongst our early hominin ancestors has long been a topic of discussion plagued with uncertainties and, as a result, divergent anthropological models (Moro Abadía & González Morales, 2010). In essence, even the definition of cognition has yet to be concretely defined (Benjafield et al., 2010). To some, cognition refers to one's ability to process incoming information in a conscious or unconscious manner that allows for the formation of an appropriate response (Izard et al., 1990). A slight variation from this previous definition comes from the Oxford English dictionary which defines cognition as both an action in itself and as a category of mental functioning (Benjafield et al., 2010). As an action, cognition is the process of evaluating incoming information and its further consolidation (Benjafield et al., 2010). Furthermore, this definition identifies cognition to be one of the three fundamental subsets, or faculties, of higher-order mental functioning: cognition, emotion, and volition (the ability to choose) (Benjafield et al., 2010). In line with the previous definitions, modern-day cognitive psychologists involve themselves in investigating a wide range of mental processes including intelligence, awareness, comprehension, intuition, and other higher-order functions (Benjafield et al., 2010). Furthermore, modern society is grounded in these sophisticated behaviors made possible by cognitive abilities (Izard et al., 1990). Overall, the importance of cognition is well respected and relevant among the work of modern psychologists, anthropologists, and other scientists alike (Izard et al., 1990).

However, despite the wealth of information the scientific community has obtained on cognitive processes presently, little is concretely established about the origin of modern human behavior. Driven by their curiosity to unearth more about our evolutionary history, researchers have long pondered the events leading to cognitive development (Moro Abadía & González Morales, 2010). In this chapter, a few of the proposed models of the origin of modern behavior and the involvement of Neanderthals in this puzzle will be investigated.

The European Upper Paleolithic Cultural Revolution

Of particular interest to paleoanthropologists is the apparent shift of behavior and culture occurring within the hominin species in the European Middle to Upper Paleolithic transition (Mellars, 2005). This transition period is characterized by the migration of anatomically modern humans (AMH) to Europe where they began to coexist with the Neanderthals (Rendu et al., 2019). A prominent feature of this time period is the rapid and widespread appearance of items implying the emergence of higher cognitive abilities within European populations (Mellars, 2005). For instance, the discoveries of modified animal teeth, shells, and personal decorations from this time period suggest the presence of technological and sophisticated symbolic behavior (Mellars, 2005). More notably, this period is marked by the emergence of a more refined style of artwork (Mellars, 2005). Important artwork discovered from this millennium include depictions of human sexuality in animal-human figures in southern Germany and the wall paintings of Chauvet Cave (Mellars, 2005). Consequently, this rapid burst of creative works has prompted many prehistorians to identify this period of a cultural revolution (Mellars, 2005).

To begin the discussion on the origin of modern behavior, it is fundamental that one comprehends how the modernity of a species is anthropologically identified. It is conceivable to presume that the presence of specific cranial anatomical structures as, or nearly as, sophisticated as modern day humans would be an indicator of modernity. Nonetheless, anthropologists and other researchers alike have been unable to reduce cognitive function to the presence, or absence, of any one given cranial anatomical trait (Moro Abadía & González Morales, 2010). Thus, the identification of cognition's onset in early hominin species has been problematic and cannot be solved via the fossil record alone (Moro Abadía & González Morales, 2010). Instead, scientists have looked for empirical evidence indicating the presence of modern behavioral traits and practices that would imply higher level cognitive abilities: burials, tool manufacturing, symbolic artwork, effective usage of large-mammals, and societal and economic organization (Henshilwood & Marean, 2003). By looking for the tangible outcomes of modern behavior, researchers have begun to get closer to understanding the evolutionary history of hominins (Henshilwood & Marean, 2003).

As mentioned prior, the origin of cognition and modern behavior has not been unanimously agreed upon. Instead, researchers have come to propose three main hypotheses regarding cognitive development: the 'Independent-Evolution Mode' or 'Multiple Species Model' (MSM), the 'Single Species Model' (SSM), and the Later Upper Pleistocene Model (Henshilwood & Marean, 2003; Moro Abadía & González Morales, 2010).

The Multiple-Species Model (MSM)

The 'Multiple-Species Model' (MSM) proposes that the evolution of cognitive development was not exclusively limited to the AMH of African origin (Moro Abadía & González Morales, 2010). Grounded in the principles of Darwinian genetics, researchers propose that modern behaviors arose due to natural selection within the local European Neanderthal populations (Moro

Abadía & González Morales, 2010). In this model, the external climate and environmental conditions in which the Neanderthals were subjected to were the selective pressures (Moro Abadía & González Morales, 2010). It is proposed that the rapid environmental changes occurring from 60,000 to 30,000 before present (BP) prompted the displacement of previously settled neanderthal populations (Moro Abadía & González Morales, 2010). Uprooted from their native geographic locations, Neanderthal populations faced insecurity in food availability and safety (Moro Abadía & González Morales, 2010). Due to these pressures, conflict between neighbouring populations elevated as resources and suitable spaces for habituation became scarce (Moro Abadía & González Morales, 2010).

Speaking in terms of the basic Darwinian principles, these circumstances became effective selective pressures for the Neanderthal populations (Moro Abadía & González Morales, 2010). In light of these novel pressures, symbolic communication and other behaviors deemed modern (i.e. valuing technological knowledge, implementing economic systems, etc.) obtained a high adaptive value (Moro Abadía & González Morales, 2010). Adaptive value refers to a behavior's ability to contribute to the overall well-being and reproductive potential, or fitness, of a population (Cohen & Axelrod, 1984). As a result, neanderthal populations who had embodied these behaviors were selected for—they had greater reproductive success in their new environment (Moro Abadía & González Morales, 2010). In contrast, populations that did not integrate these behaviors into their way of life faced hardships and, eventually, died off (Moro Abadía & González Morales, 2010). According to this model, this led to the propagation of these modern behaviors among the future Neanderthal descendents (Moro Abadía & González Morales, 2010).

Observed as a vital source of evidence for this model is the Grotte du Renne, a cave located in France (Moro Abadía & González Morales, 2010). At this location, the remains of Neanderthals have been correlated to bone tools, personal ornaments, among other items indicative of modern behaviors

(Caron et al., 2011). However, it is unclear whether the use of these symbolic and technological items evolved independently in the Neanderthal species or was induced through the mimicry of AMH following their migration to Europe (Caron et al., 2011).

The 'Single-Species Model' (SSM)

In the 'Single-Species Model' (SSM), anthropologists propose the modern behavioral characteristics fundamental to the European Upper Paleolithic revolution (i.e. symbolic artwork, technological advances, etc.) antecedently emerged in Africa (Mellars, 2005). They propose that the origin of modern behavior lies in the African Middle Stone Age thousands of years before its appearance in Europe (Mellars, 2005). Among the evidence for this proposal is the discovery of bone tools, geometrical designs in ochre at Blombos, and other items involving symbolic expressions (Mellars, 2005). All of these findings mirror the key aspects of the 'cultural revolution' in the Upper Paleolithic (Mellars, 2005). The wide range of archeological findings have now pointed to the presence of symbolic behavior in African AMH 80,000 to 90,000 BP (Mellars, 2005). However, as the definition of cognition remains inconclusive, the conclusions drawn from these findings on the cognitive abilities of AMH are still controversial. Yet, these striking findings of the archeological record should not be easily dismissed. Nonetheless, supporters of this model postulate that the modern behavior observed within the Neanderthal species was due to observational learning, not independent origin (Mellars, 2005). They propose that, upon the migration of AMH north-east, the intermingling of the two species resulted in a cultural diffusion (Mellars, 2005).

The Later Upper Pleistocene Model

The Later Upper Pleistocene model may be seen as a derivative of the SSM. In this model, anthropologists hypothesize that any form of modern human behavior was not present until some point after 50,000 years ago (Henshilwood

& Marean, 2003). As such, in contrast to the SSM, those supporting this model disagree with the presence of modern behaviors in the African Middle Stone Age or Middle Paleolithic era (Henshilwood & Marean, 2003). Instead, they classify these time periods to be characterized by primitive material culture, low effectivity in food collection (i.e. limited knowledge of seasonal implications on food supply, no large-animal hunting, etc.), and the absence of symbolic art (Henshilwood & Marean, 2003). Evidence collected suggesting the presence of these behaviors is dismissed due to lack of contextual considerations (Henshilwood & Marean, 2003). For example, items that modern-day anthropologists may identify as tools are claimed to have not been used in a sophisticated and cognitively-aware manner (Henshilwood & Marean, 2003). Proponents of this model argue that researchers have taken a too modernized anthropological approach in the identification of these items (Henshilwood & Marean, 2003).

The boundary separating the Neanderthals from the anatomically modern humans, located in Eastern Mediterranean region of Eastern Asia, fluctuated with the change in environmental conditions (Henshilwood & Marean, 2003). To elaborate, as the weather warmed, modern humans would move northward to occupy the territory previously inhabited by the Neanderthals (Henshilwood & Marean, 2003). In a likewise manner, in colder, harsher weather, the Neanderthals would push the boundary south-eastern, expanding their own land capacity (Henshilwood & Marean, 2003). This thermo-dependent equilibrium of the populations is thought to have been maintained until approximately 40,000 years ago (Henshilwood & Marean, 2003). At this point, proponents of this theory propose modern humans had developed the behaviors necessary for survival in colder habitats; consequently, they began to migrate northward into Europe (Henshilwood & Marean, 2003).

In explaining this sudden migration, the Later Upper Pleistocene Model postulates that the development of modern behaviors among the AMH was fundamental (Henshilwood & Marean, 2003). Following this event, the Neanderthal populations became extinct in the next 12,000 years (Henshilwood & Marean,

2003). To those supporting this model, this extinction serves as evidence that the Neanderthals were outcompeted by the cognitively superior AMH, eventually leading to their demise (Henshilwood & Marean, 2003).

Which Model is Correct?

The answer to this question is inconclusive—each model is sustained by its probable truths while at the same time being subjected to its own criticisms. Overall, the SSM and Later Upper Paleolithic model clearly contradict and identify fundamental flaws within the MSM (Henshilwood & Marean, 2003; Mellars, 2005). First, following the migratory path of AMH, it has been determined that they had originated in Africa and migrated north towards Europe (Mellars, 2005). This claim is confirmed by several pieces of evidence including mitochondrial DNA research, the collected fossil records, and studies of the Y chromosome (Mellars, 2005). As noted previously, the time period in which this migration occurred was parallel to the first appearances of technological advancements in Europe (Mellars, 2005). To many paleoanthropologists this fact cannot, and should not, be overlooked (Mellars, 2005).

Furthermore, critics of this hypothesis have compared the archeological record in the Middle Stone Age of Africa to Upper Paleolithic Europe to disprove the theory (Mellars, 2005). Specifically, the slower appearance of innovations in Africa contrast the rapid instillation of technological advancements that occurred within Europe (Mellars, 2005). Moreover, as mentioned prior, evidence of 'modern human' behavior was found in Africa 30,000 to 40,000 years earlier to its integration into populations located in Europe (Mellars, 2005). Thus, proponents of these models highlight that the rapid proliferation of modern behaviors throughout Europe was more likely the result of cultural diffusion between AMH and Neanderthals than any other source (Mellars, 2005). Overall, critics argue the MSM does not adequately account for the rapid proliferation of innovation across Europe nor the potential impact of more technologically advanced populations migrating to Europe (Mellars, 2005).

Finally, critics of the MSM point out the complications with proposing that the Neanderthal populations were as cognitively competent as the AMH (Mellars, 2005). If this were true, the natural selection of the AMH of African origin over the Neanderthals becomes complicated. As the point in time in which AMH began to migrate through Europe, Neanderthals had already been European inhabitants for over 200,000 years (Mellars, 2005). Due to the driving forces of natural selection, over the course of this time period Neanderthals had become biologically adapted to their surrounding environmental conditions (Mellars, 2005). For instance, researchers have hypothesized that their large chest cavities, among other traits, may have allowed them to be more suited for the harsh cold or high levels of excursion (Churchill, 2006). Originating in Africa, AMH did not have these types of physiological modifications and would have been maladapted to endure cold weather conditions (Mellars, 2005). Therefore, if the neanderthal populations were cognitively equivalent to their AMH counterparts, their physiological adaptations should have donned them a higher overall reproductive fitness (Mellars, 2005). Yet, Neanderthal populations became extinct within the time period of 30,000 - 40,000 BP, coinciding with the AMH European migration (Horan et al., 2005). While the specific underlying causes of the extinction continue to be the source of scientific debate, the potential in a variance in cognition and implementation of modern behaviors within the two species cannot be overlooked.

Nevertheless, the possibility of the MSM holding more truth also should not be overlooked. It is possible that Neanderthals had developed modern behaviors but expressed them in different ways than the AMH. In the past, the discussion of whether Neanderthals were capable of symbolic thinking has previously been clouded by scientists' narrow perspective of artwork (Moro Abadía & González Morales, 2010). In general, Paleolithic art can be classified into two main categories: rock art and portable art (personal ornaments) (Moro Abadía & González Morales, 2010). Until recently, heavy emphasis had been placed on the significance of cave art, its presence indicative of higher cognitive abilities (Moro Abadía & González Morales, 2010). In contrast, portable art had vastly been viewed as secondary in importance (Moro Abadía & González Morales, 2010). Instead, the ornaments that had been correlated to Neanderthal populations were seen to be for beauty purposes, nothing more (Moro Abadía & González Morales, 2010). In this way, the scientific community undermined the importance of these items in the archeological record. However, this perspective has changed in recent years (Moro Abadía & González Morales, 2010). Now, these ornaments are seen as a symbolic representation of personal and ethnic identity (Moro Abadía & González Morales, 2010). Furthermore, anthropologists now appreciate that the intricate movements necessary to craft such ornaments are comparable to those essential for cave paintings (Moro Abadía & González Morales, 2010). With this redefinition of art and newfound appreciation of portable art it has become necessary to reevaluate our current archeological collection. Perhaps, in doing so, the Neanderthal's cognitive abilities will be re-defined in a new more comprehensive light.

Final Thoughts

In summary, if symbolic art did not originate in the Neanderthal populations it does not appear to be resultant of their overall cognitive ineptness (Moro Abadía & González Morales, 2010). Through the archeological record, Neanderthals appear to have been capable of symbolic artwork and tool manufacturing (Mellars, 2005). It seems, at the very least, it can be concluded that Neanderthal populations did contain some aspects of modern behavior (Mellars, 2005). However, the debate remains as to how these behaviors became integrated within their populations. Possibly, the beginning of modern behaviors in Europe came about following the migration of AMH from Africa (Henshilwood & Marean, 2003). Such is postulated by both the Later Upper Paleolithic model and SSM (Henshilwood & Marean, 2003; Moro Abadía & González Morales, 2010). On the other hand, perhaps Neanderthals did not attain modern behaviors through cultural diffusion but rather via intra-species natural selection (Mellars, 2005). These ideas will continue to be central to anthropological discussions for years to come. In fact, it is possible that the origin of modern behavior in Europe may have followed a drastically different scheme than the ones previously discussed. As anthropological methods continue to advance, it is possible that one day researchers may unearth the answers to the mysteries of modern behavior's evolution.

Chapter 5
Neanderthal Cranio-Facial Morphology
by Ann Ping

Introduction

Homo neanderthalensis, or the Neanderthals, was a hominin species that inhabited Europe and Western Asia between around 200 000 and 30 000 years ago (Bailey, 2002). Its cranio-facial morphology is of special interest, especially in regards to its evolution (Clement et al., 2012). Some distinguishing features of the Neanderthal cranio-facial morphology include rounded and laterally projecting parietal bones (skull bones that form the sides and roof of the cranium), facial prognathism (bulging out of lower jaw), a wide and tall nasal aperture, a wide and projecting nasal bridge, a depressed nasal floor, and large anterior tooth crowns (the revealed part of the tooth) relative to posterior tooth crowns (Clement et al., 2012; Weaver, 2009). There are multiple hypotheses that address the evolution of the Neanderthal cranio-facial morphology, including adaptation to high masticatory (the use of the teeth for food) and paramasticatory (the use of the teeth for tools or other non-food related uses) forces, adaptation to cold climates, and genetic drift (Clement et al., 2012). Genetic drift is when allele frequencies in a population change due to random chance. Additionally, the relation of Neanderthals to modern humans is still a topic debated by anthropologists to this day (Bailey, 2002). The two competing models for modern human origins are the Recent African Model and the

Multiregional Evolution Model (Bailey, 2002). The Recent African Model states that Neanderthals contributed little or nothing to the human gene pool because Neanderthals were replaced by emigrating African modern humans (Bailey, 2002). In contrast, the Multiregional Evolution Model posits that Neanderthals and other archaic humans played a significant role in modern human origins through gene flow, selection, and genetic drift (Bailey, 2002; Relethford, 2008). Comparisons between Neanderthal cranio-facial morphology, especially dental morphology, and modern human cranio-facial morphology shed light on the evolutionary relationship between Neanderthals and modern humans (Bailey, 2002). Hence, this chapter will review the research on Neanderthal cranio-facial morphology in reference to its own evolution as well as to the evolution of modern humans.

The Neanderthal Dentition

Among many observations of Neanderthal dental morphology, a prominent one is the large amount of wear in the anterior (front) teeth relative to the posterior (back) teeth (Weaver, 2009). This disparity in tooth wear may have important implications in the evolution of the Neanderthal's unique facial anatomy (Clement et al., 2012). The anterior dental loading hypothesis predicts that key features of the Neanderthal's cranio-facial morphology, such as their mid-facial prognathism, greater upper facial height, and swept-back cheekbones may have been adaptations to reduce the high mechanical loads on their anterior teeth (Weaver, 2009; Clement et al., 2012). In fact, not only is there greater wear on the anterior teeth, but the anterior teeth are also larger than the posterior teeth, and they have bulging, robust crowns (Clement et al., 2012). These features are believed to have provided greater buttressing for the crown and a greater volume of tooth which confers resistance to wear (Clement et al., 2012). It has also been found that the prognathic mid-facial profile of Neanderthals would have offered greater resistance to high torsional (rotational) stresses induced by anterior dental loading (Demes, 1987). This is because greater anterior loads increase the bite force on one side of the transverse axis, producing heavy rotational forces in the upwards and posterior direction (Demes, 1987). Meanwhile, the unique positioning of the infraorbital facial plates (bony plates of the cheek underneath the eye) renders them more efficient in opposing the high torsional forces generated by high anterior dental loads (Demes, 1987). In other words, some aspects of the Neanderthal cranium structure appear to oppose the forces placed on the anterior teeth. Additionally, the anterior positioning of the masticatory muscles, the posterior positioning of the incisors, and the anterior positioning of the molars have been shown to be mechanically favourable in producing strong anterior bite forces (Spencer & Demes, 1993). Hence, such biomechanical evidence lends support to the anterior dental loading hypothesis, showing that Neanderthal cranio-facial morphology is in part specialized for intensive anterior tooth use (Spencer & Dames, 1993).

However, the research is divided on whether the anterior dental loading hypothesis is adequately supported. Some paleoanthropologists have questioned a critical assumption of the anterior dental loading hypothesis, which is that the Neanderthal masticatory system was able to generate such heavy loads in the first place (O'Connor et al., 2005). Further research shows that the amount of incisal bite force the Neanderthals were able to generate was not significantly greater than that of modern humans (O'Connor et al., 2005). Moreover, O'Connor et al. (2005) assert that the masticatory muscles (the muscles used to chew food) were less efficient than those of humans in producing large forces. In this case, efficiency is quantified as the ratio of bite force to muscle force. In fact, studies show that the typical facial features associated with high bite forces are not consistent with those of Neanderthals (O'Connor et al., 2005). For instance, large bite forces are generally produced by large faces, not necessarily the high faces which Neanderthals exhibit (O'Connor et al., 2005). Given these results, some paleoanthropologists propose an alternative to the anterior dental loading hypothesis. They hypothesize that the high levels of anterior tooth wear may have been associated with repetitive anterior tooth use rather than forceful anterior tooth use (O'Connor et al., 2005).

At this point, one may be wondering what the Neanderthals were doing to cause such high amounts of anterior tooth wear. Regarding dietary-related uses, it is suggested that Neanderthals used their anterior dentition to clamp down on tough meat before slicing it near their mouths with a stone tool (Krueger et al., 2019). Neanderthals may also have been eating coarser and more abrasive foods (Krueger et al., 2019). There is also evidence that Neanderthal tooth wear may have non-dietary related causes, such as frequent probing by toothpick-like tools (Frayer & Russell, 1987). Neanderthals also used their teeth in tool production, hide preparation, wood softening, and weaving tasks (Krueger et al., 2019).

Teeth are especially useful in phylogenetic studies for two main reasons. Firstly, teeth contain morphological information that is controlled to a greater extent by genes than are skeletal features (Bailey, 2002). Secondly, teeth are abundant and well-preserved in the fossil record (Bailey, 2002). Hence, the dentitions of Neanderthals and modern humans shed a lot of light on the evolutionary relationships between Neanderthals and modern humans (Bailey, 2002). Bailey (2002) finds that a high frequency of Neanderthal postcanine teeth are marked by complex occlusal surface topography and an asymmetrical lingual contour. In other words, many of the studied Neanderthal tooth samples demonstrated complex patterns at the interface where the upper and lower teeth meet and asymmetry on the side of the tooth that is adjacent to the tongue. This is in contrast to modern humans, who frequently exhibit a greatly simplified occlusal surface topography and a symmetrical lingual contour (Bailey, 2002). Additionally, Neanderthal teeth exhibit a strong and continuous transverse crest (a type of ridge on the surface of the tooth) and multiple lingual cusps (elevations of the tooth located near the tongue) (Bailey, 2002). Yet again, this is in contrast to modern humans, who lack a well-developed continuous transverse crest and less frequently exhibit multiple lingual cusps (Bailey, 2002). These disparities suggest that Neanderthals exhibit a unique dental morphology relative to modern humans. However, to obtain a better understanding of their evolutionary relationship, one must determine whether the Neanderthal dental pattern is primitive or derived (Bailey, 2002). Primitive features are inherited from the common ancestor of a clade, whereas derived (or autapomorphic) features are unique to a given clade. To determine whether the Neanderthal dental pattern is primitive, it is compared to that of Homo erectus, a closely related ancestor of Neanderthals (Bailey, 2002). It turns out that H. erectus lacks the unique Neanderthal dental morphology described above. The teeth of H. erectus either lack a transverse crest or exhibit an interrupted transverse crest and they have a symmetrical lingual contour (Bailey, 2002). These traits are remarkably similar to those of modern humans. Therefore, assuming that H. erectus represents the primitive conditions, Neanderthal dental characteristics are not primitive, but rather derived. Moreover, because the dental morphology of modern humans is consistent with that of H. erectus, modern humans seem to retain these primitive dental characteristics (Bailey, 2002). This investigation shows that Neanderthals may have evolved separately from modern humans for a long time (Bailey, 2002). Therefore, because Neanderthals seem to have played little role in the evolution of modern humans, this investigation provides evidence against the Multiregional Evolution Model. Additionally, because derived features can point to directional natural selection (when one extreme phenotype is favoured over the others) or genetic drift (Weaver, 2009), this may also have been the case with Neanderthals.

Overall, a review of the Neanderthal dentition showed that the anterior dental loading hypothesis may not be true. This indicates that adaptations to reduce high levels of force on the anterior teeth may not have influenced the evolution of the Neanderthal cranio-facial form. Additionally, a comparison between the dentitions of Neanderthals, humans, and H. erectus shows that Neanderthals likely played little to no role in the evolution of modern humans.
Neanderthal Nasal Morphology

Another hypothesis regarding Neanderthal cranio-facial evolution is known as the nasal radiator hypothesis, which predicts that mid-facial prognathism and a wide nasal cavity were adaptations to warm inspired air (Clement et al., 2012).

The warming of inspired air would have reduced the effects of cold climates on the brain (Clement et al., 2012). Additionally, it was assumed that the large dimensions of the external nose, the depressed nasal floor, and large paranasal sinuses (air-filled spaces in the nasal cavity) were adaptations to add moisture to inspired air and to dissipate heat in cold and dry climates (Clement et al., 2012). Despite initial support for the theory in the 1950s and the 1960s, more recent research has questioned its legitimacy. While some aspects of the Neanderthal nose are consistent with cold climate adaptation, such as internal nasal size and shape and interorbital distance (the distance on the skull between the eyes), some paleoanthropologists argue that the characteristic wide nasal aperture of Neanderthals is strongly inconsistent with cold climate adaptation (Holton & Franciscus, 2008). In fact, wide nasal apertures are more consistent with tropical climates (Holton & Franciscus, 2008). There's also evidence that narrow nasal apertures are more common among modern humans who live in higher latitudes when compared to modern humans who live near the equator (Holton & Franciscus, 2008). This gives rise to an important question: why do Neanderthals, who lived in colder and drier climates, exhibit large nasal apertures?

One explanation for the Neanderthals' wide noses is that their nose morphology was constrained by their large anterior dentition and the increased growth of the anterior palate (roof of the mouth), which would have pulled apart the edges of the nasal aperture (Holton & Franciscus, 2008). However, this hypothesis stands shakily on the assumption that the anterior palate and dentition constrain nasal dimensions rather than the reverse. Holton and Franciscus (2008) tested this hypothesis and subsequently rejected it after finding a weak association between intercanine breadth (the width between the two canines) and nasal breadth. Intercanine breadth is a measurement of the breadth of the anterior dentition and palate. Instead, they found that the degree of facial prognathism demonstrated a stronger association with nasal breadth. In other words, Neanderthals with a more protruded jaw also had a wider nose. These findings better explain Neanderthal nose morphology. Moreover, because facial prognathism is unrelated to the climate, not only is the nasal radiator hypothesis rejected by these findings, but any sort of climate adaptation acting as a selecting force in Neanderthal cranio-facial morphology has become less likely.

Genetic Drift

Thus far, two main hypotheses explaining the evolution of Neanderthal cranio-facial morphology, the anterior dental loading hypothesis and the nasal radiator hypothesis, were shown to be weakly supported by evidence. So what could have driven Neanderthal cranio-facial evolution? Many paleoanthropologists agree that genetic drift and local selection in isolated Neanderthal populations are the most likely mechanisms (Weaver, 2009; Clement et al., 2012, Franciscus, 2003). This is because geographic isolation in glacial Europe would have provided the optimum conditions for genetic drift (Clement et al., 2012). Additionally, recurring climatic crises during the time would have reduced available territory, resulting in population bottlenecks (drastic reductions in population) and subsequent genetic drift (Clement et al., 2012). Franciscus (2003) proposes that local selection was secondary to the stochastic events that lead to genetic drift. This explanation reconciles the fact that there is still some substance, albeit little, to anterior forces and climatic adaptations as factors that may have influenced Neanderthal cranio-facial evolution.

Conclusion

This chapter highlighted hypotheses for the evolution of Neanderthal cranio-facial form, concluding that the most likely mechanism for evolution was genetic drift, supported secondarily by local selection. The evolutionary relationship between Neanderthals and modern humans was reviewed as well, leading to the conclusion that Neanderthals played little to no role in the origin of modern humans. Finally, although the insights that the Neanderthal face and cranium provide us are powerful, it is important to consider that the features examined in this chapter compose merely a fraction of the entire Neanderthal morphology. Accounting for postcranial Neanderthal morphology may shed new light on the conclusions previously reached. Ultimately, the fossil record is an invaluable tool in investigating evolutionary and phylogenetic relationships; it is thanks to the fossil record and passionate paleoanthropologists that hominid species that lived hundreds of thousands of years ago can be studied in such detail in the 20th and 21st century.

Chapter 6
Ethical Principles of Genetic Research
by Ariana Balassone

Introduction—Ethics and its Importance

Ethics is a branch of philosophy that governs moral principles and involves suggesting ways to correct immoral activity. It is important to discuss the ethical issues involved when researching human origins because with the advancement of technology, the new methods of genetic testing may be somewhat immoral in some eyes. The difficulty in examining ethics is defining morality. Different individuals have varying interpretations and perspectives of what is considered acceptable in any field of research.

Before ethics is examined, it is necessary to discuss the existing methods of genetic testing in the field of human evolution and origins. Natural selection is defined as the particular survival and reproduction of individuals on the basis of their genotype. The environment plays a key role in evolution because it creates the selective pressures that define one's fitness, which is the ability of an organism to survive and reproduce in a certain environment. To study natural selection, researchers need to examine genetic diversity across the world, as well as the DNA of related hominids, such as Neanderthals (Vitti et al., 2012). As genetic technologies advance, the responsibility for such tools increases. There are immediate issues involved in genetic analysis, such as consent of the individuals participating in the research study and if the information discovered can be utilized for purposes other than the original study (Vitti et al., 2012). On a broader scale, there are also issues relating society to genetics, such as conflicts between personal choice and public health that can emerge from the development of genetic enhancements (Vitti et al., 2012). Evolutionary genomics has opened an entirely new discussion which presents new and unaddressed ethical issues.

Evolutionary Genomics

Evolutionary genomics unites the concepts of evolutionary and molecular biology, making it a powerful field of research (Vitti et al., 2012). Through methods of genetic analysis, researchers are able to apply statistics to the genome when examining genetic variation, the history of individual genes, and loci recently subject to selection (Vitti et al., 2012). However, this technology has the potential to reveal facts about the human species that can be misinterpreted by the public to fit biased views (Vitti et al., 2012). This is why it is absolutely essential that researchers communicate their results very precisely to ensure that non-scholarly sources cannot misinterpret them in an unjust way. In the past, evolutionary discoveries in genetics have been used to rationalize discriminatory policies: genetic discrimination occurs when people are treated differently due to underlying genetic mutations that causes or increases the risk of an inherited disorder (Vitti et al., 2012). For instance, a boy in California in the sixth grade had to leave school after receiving DNA test results that showed genetic markers for cystic fibrosis (Joly et al., 2013). Therefore the study of human evolution requires an ethically self aware research agenda and a discussion between biologists, bioethicists and policy-makers. There are several concepts involved in evolutionary genomics, such as ancestry and genetic disorders that can be easily misinterpreted (Vitti et al., 2012). How can researchers and science journalists avoid these miscommunications? Current research progress and future directions involving evolutionary genomics will be discussed throughout this chapter.

Genomic Analysis From a Scientific Perspective

Recall that one of the primary causes of evolution is natural selection. Over the years humans have been exposed to new environmental conditions and diseases; those who possessed better adaptations to the changing surroundings survived and passed on their alleles with a higher frequency (Bamshad & Wooding, 2003). The process of natural selection has left marks in our genome which act as identifiers of evolution that might underlie disease variation and drug metabolism (Bamshad & Wooding, 2003).

Humans differ from each other in numerous ways including physical appearance, susceptibility to disease, and symptom severity. Researchers can identify functionally important genes by identifying loci that have been acted on by natural selection (Bamshad & Wooding, 2003). Variants that increase the fitness of an individual will increase in frequency as a result of positive selection. The favourable allele will rise in the population until it reaches fixation, and when this happens the loci under selection will evolve with the linked surrounding region of the genome (Vitti et al., 2012). Positive selection leaves signatures in the genome that can be detected through estimated calculations.

Indicators of Selection in the Genome

Selective sweep events (when a favourable mutation increases in frequency until it becomes fixed in a population) can reduce diversity in a region because one allele will rise to a high frequency and simultaneously increase the frequency of neighbouring alleles (Biswas & Akey, 2006). This increases homogeneity across populations and can be identified with long strings of DNA sequence that contain small amounts of single nucleotide polymorphisms (SNPs) (Biswas & Akey, 2006).

Recall that the allele under selection tends to take with it the surrounding genes, which means that new variants which neighbour the allele under selection will rise in frequency as well. These new alleles are known as "derived alleles" because it takes a long time for them to rise prevalently in the absence of selection taking place directly on them (Biswas & Akey, 2006). Thus, an excess of derived alleles in a region is indicative of positive selection.

As an allele is passed down through generations, it has the opportunity to undergo random recombination events that cease its association with other alleles (its haplotype). Essentially, this means that the allele is linked to its surrounding region (Biswas & Akey, 2006). If selection occurs rapidly, such recombination events are more unlikely to happen, and regions of DNA with longer than usual haplotypes are also signatures of positive selection (Biswas & Akey, 2006). Furthermore, comparatively large differences in allele frequencies between populations is further indication of selection as this would be expected with different conditions and hence different adaptations (Biswas & Akey, 2006). Thus, selection occurs in one population but not the other.

With new advances in technology, new methods of genetic analysis that are more efficient and accurate are more readily available. Researchers no longer have to rely solely on the archeological record for evidence of selection (Vitti et al., 2012). They can now use these new technologies to analyze DNA from humans across the world and determine which loci have likely undergone selection. These technological advancements also permit the use of high-throughput methods (analyzing dozens of samples at once), rather than a case by case analysis, saving much time (Vitti et al., 2012). There are evidently many benefits with using new and improved technologies to analyze genomes from a scientific perspective. One particular method of genomic analysis is whole genome scans, which determine the entire genetic sequence of an organism (Vitti et al., 2012). In doing this, researchers are able to isolate genetic markers that are remnants of selection taking place, and these markers can be used to reveal the loci that was under selection for further investigation (Vitti et al., 2012).

The primary use of genome wide scans was in creating maps of human variation. HapMap Consortium and the 1000 Genomes Project have made genetic information easily accessible and available, providing essential information for human evolutionary genomics to progress (Vitti et al., 2012).

Through new and innovative technologies, evolutionary genomics has progressed tremendously. Researchers have since identified novel adaptations for lactase persistence in Europe, verifying their predictions that the lactase enzyme originated in Europe due to the domestication of cattle (Vitti et al., 2012). They have also examined genes for skin pigmentation in Asia and Europe, ear wax and hair production, the development of teeth and production of sweat in Asia, and resistance to infectious diseases in Africa (Vitti et al., 2012). Additionally, a family of genes has been discovered in the Tibetan populations that is under selection and may help the body adjust to changes in high climate (Yi et al., 2010). These adaptations further signify the importance of climate, diet, and infectious diseases in propelling evolution. These discoveries also show the importance of evolutionary genomics to researchers.

Although these accomplishments are encouraging, there are still many genes that need to be functionally identified before a detailed phenotypic, molecular, and mechanistic characterization of population genetics is established (Vitti et al., 2012). Current computational methods are best for new mutations that occur at a high frequency, whereas statistical tests are being developed for mutations for weaker selection (Vitti et al., 2012). Furthermore, current methods are used to identify single loci mutations, although many mutations in response to adaptations are associated with large gene complexes (Vitti et al., 2012). It is also important to identify epistatic interactions between genes, as epistasis is involved in evolution (Vitti et al., 2012). Researchers should keep this in mind as current methods continue to be refined. Moreover, there are many other processes that contribute to evolution, including genetic drift, expansion, and population bottlenecks (Vitti et al., 2012). Thus, new methods need to be developed that fit these models of evolution, rather than solely relying on the selective sweep model. Nevertheless, the selective sweep model of evolution has inspired many methods involved in evolutionary genomics and it presents a favourable opportunity to examine ethical issues while technologies are currently being developed.

Ethical Issues in Evolutionary Genomics

Evolutionary discoveries are often warped to fit prejudicial opinions. This is the reason for which it is important that researchers carefully and precisely word their results and conclusions, to avoid miscommunication with the general public. First, evolution can be a sensitive and misunderstood topic by the public itself. The idea of 'survival of the fittest' can often be misinterpreted as one race being superior to the other. For instance, in 1864 Herbert Spencer described survival of the fittest as the "preservation of favoured races in the struggle for life" (Spencer H, 1865, p.531 - 550). While it is now acknowledged that natural selection does not favour certain ethnic groups, this innate misreading has been used to justify beliefs such as social Darwinism, in which welfare policies are restrained to eliminate the 'socially unfit' (Vitti et al., 2012). Furthermore, the idea that evolution is progressive has given birth to the false notion that modern African populations are ancestral, and thus less evolved than other racial groups (Vitti et al., 2012). These types of misconceptions invite racism and political bias into society, which is an unfortunate consequence that has followed evolutionary research. Therefore, researchers must be aware of how the public can react and twist such evolutionary discoveries when releasing results (Vitti et al., 2012).

Researchers must help the audience avoid misconceptions in genetic determinism, which is the belief that the genes are directly responsible for all aspects of human behavior (Vitti et al., 2012). Such misunderstandings have been used to justify abolishing initiatives such as the Head Start Program in the United States. This program helps children from low income families obtain an education and improves social and cognitive development (Vitti et al., 2012).

However, discoveries in genetic variation involved in cognitive factors have been twisted by the public to argue that cognitive abilities are not influenced by environmental factors. Nothing known about genetics supports these ideologies, however they can be convincing arguments and used in the wrong ways for discoveries in evolutionary genomics.

Taking into account the dark past of publishing evolutionary discoveries, researchers must be very careful in wording their results such that nothing can be twisted to fit biased agendas. Moreover, researchers may discover actual differences in the genes responsible for cognitive and behavioural trails between populations which could evidently create problems and controversy. For instance, recent publications suggested natural selection acting on two sets of genes; microcephalin (MCPH1) and abnormal spindle-like microcephaly associated (ASPM), in primarily Euroasian groups but not West African populations (Vitti et al., 2012). Due to the fact that mutations in these genes are associated with microcephaly and static mental delay, the authors proposed that these genes were involved in brain size and that they may have recently undergone selection (Vitti et al., 2012). This inspired racist conclusions on one hand and opposition from the other. Since the publication of these results, recent studies have not confirmed that the genes are involved in brain size nor that they underwent selection recently (Vitti et al., 2012). However, the reaction of the public further demonstrated the controversial idea of evolutionary genomics and the requirement for scientific discretion when discussing conclusions. Researchers may discover differences in genes associated with cognitive and behavioural traits between populations. Currently, there are no such cases, however researchers are continuously investigating the evolution of genes involved in cognitive development and they could uncover some new information (Vitti et al., 2012). Therefore methods of careful communication between researchers and the public need to be discussed in the case that controversial information is discovered. Researchers should be attentive in three extensive areas: the distribution of results, in their own techniques, and in their role controlling public talk of scientific topics (Vitti et al., 2012).

Chapter 7
Evolutionary Relationship Between Humans and Neanderthals
by Romina Tabesh

Introduction

Humans have many distinct characteristics; these include unique cognitive and physical aspects in comparison to other primates. Comparative genomics and analyses of the human and chimpanzee genomes continue to reveal information about modern human traits (Noonan, 2010). This field of study involves biological research in which different organisms are compared on the level of genomic features, such as genes, gene order, DNA sequence, regulatory sequences, as well as other structural landmarks within the gene. Comparative genomics has allowed geneticists to study changes in the sequence of the human genome which possibly contributed to the evolution of modern human traits. This is especially important because it allows geneticists to better understand human evolution, as well as the physical and cognitive abilities of modern humans in comparison to other primates.

Background

Hominins is a term given to a group of species which includes modern humans and their ancestors. Hominins have existed for approximately 4 million years, starting with Australopithecines. The modern human species, Homo sapiens, however, only made an appearance about 200,000 years ago. Homo sapiens are not the sole representative species of Hominids that are now known as modern humans. Interestingly, paleontologists discovered that early Homo sapiens may have shared their landscape with other Hominin species; for example,t Neanderthals overlapped with early Homo sapiens in vast areas of Europe (Noonan, 2010). Newly discovered evidence also suggests that until about 40,000 years ago, when the Neanderthals disappeared, ancient humans and Neanderthals lived very closely with one another—close enough to interbreed (Noonan, 2010). In addition to studying fossils and artifacts such as art and tools, scientists have discovered methods for extracting DNA from old fossils (Noonan, 2010). Metagenomics, which is the study of genetic material obtained directly from environmental samples, has allowed scientists to study similarities between modern humans and Neanderthals and has made it possible to approximate the time at which modern humans and Neanderthals last shared a common ancestor (Nooan et al., 2006).

Prior to the use of metagenomics, knowledge of Neanderthals was limited and based on only a small number of remains and artifacts, leaving little evidence that could be used to make inferences Neanderthal physiology, behaviour, and their relationship to modern humans (Nooan et al., 2006). Scientific advancements, however, now enable the use of a developed Neanderthal metagenomic library which enables geneticists to sequence and analyse Neanderthal genomes in much greater detail. The use of this technology has also enabled geneticists to calculate human-Neanderthal divergence. It is estimated that the two species last shared a common ancestor 706,000 years ago, and that the ancestral populations of humans and Neanderthals diverged approximately 370,000 years ago, before anatomically modern humans emerged (Nooan et al., 2006). The Neanderthal metagenomic library has also been used to compare the degree of similarity between modern human and Neanderthal DNA. Moreover, it was found that the human and Neanderthal genomes are no less than 99.5% identical (Nooan et al., 2006). Such analyses and discoveries enhance understanding of the evolutionary relationship between Homo sapiens and Homo neanderthalensis, and also enable geneticists to better understand the evolution of traits possessed by modern humans.

Neanderthals are evidently the closest hominid relatives of modern humans (Nooan et al., 2006). The two species coexisted and interbred as late as 30,000 years ago in Europe and western Asia (Nooan et al., 2006). Since then, the species of modern humans has scattered, occupying areas across the Earth and far surpassing primate species of hominids in numbers, technological development, and environmental impact (Nooan et al., 2006). This success is partly due to complex biological features such as a relatively large brain size, bipedalism (the ability to walk on two legs), and modifications in physical morphology which began to emerge in human ancestors before the rise of modern humans (Noonan, 2010). These adaptations along with other, more complex ones, were a major factor facilitating the evolution of unique human behavioural traits, for example, language. These adaptations are caused by changes in the DNA sequence of the human lineage marking the human-chimpanzee split (Noonan, 2010). Identifying and analyzing such changes in genomes have become an important focus of human genetics and genomics.

Techniques Used to Study Neanderthal Genomes

Genetic analysis technologies such as the polymerase chain reaction—a method widely used to rapidly make millions to billions of copies of a specific DNA sample, allowing scientists to take a very small sample of DNA and amplify it to a large enough amount to study in detail— have also been used to amplify Neanderthal mitochondrial sequences. These sequences are unique as they indicate inheritance from only the mother and can be used to trace ancestry. They have been analysed by geneticists who discovered the most recent common ancestor of modern humans and Neanderthals lived approximately 500,000 years ago, which is well prior to the emergence of modern humans (Nooan et al., 2006). But, a limitation to this mitochondrial data obtained from Neanderthals using PCR is that it does not provide access to the gene and regulatory sequence differences between these hominids and modern humans, which would aid in revealing biological features that are unique to each species (Nooan et al., 2006).

Contemporary advances in metagenomic analysis of complex DNA mixtures, along with the introduction of high-throughput sequencing technologies now provide geneticists with methods to obtain genomic sequences from very old remains (Noonan, 2010). This was not the case previously, where direct analysis of extracts was used in an effort to recover ancient sequences (Sankararaman et al., 2014). Metagenomic libraries used by scientists now, on the other hand, allow for DNA to be immortalized. This DNA, obtained from precious ancient samples, eliminates the need for repeated destructive extractions (Sankararaman et al., 2014). In other words, the DNA obtained from fossils of a Neanderthal species can be stored in a genomic library which can be accessed and studied forever. In addition to this, once an ancient DNA fragment is cloned into this metagenomic library, it is made distinguishable from contamination which might occur during subsequent polymerase chain reaction amplification or other possible scenarios that could occur during the DNA sequencing process (Sankararaman et al., 2014).

So far in this chapter, it has been noted that Neanderthals represent an extinct lineage of hominids, existing in Europe and western Asia for approximately 400,000 years. They thrived in such regions for most of their existence but started to decline in numbers and went extinct nearly 30,000 years ago (Herrera et al., 2009). As mentioned before, their disappearance took place prior to the emergence of modern humans in those areas. This has prompted some individuals to consider whether Neanderthals were displaced by a more fit and adaptable species, being modern humans (Herrera et al., 2009). Others oppose this theory by arguing that the existence of Neanderthal genes in modern human DNA suggests the two species coexisted (Herrera et al., 2009).

The major motivation for generating a Neanderthal reference genome is to identify the extent to which modern humans differ from earlier hominids. As far as scientists are concerned, the Homo sapiens species is the only remaining human species today, hence geneticists are unable to know whether Neanderthals or our other extinct relatives shared Homo sapiens' capacity for invention, abstract reasoning, or language (Noonan, 2010). Paleontologists explore such matters based on bones and settlements, as well as artifacts left by Neanderthals. The question of similarity between Neanderthals and modern humans is of great interest to paleontologists, especially given the recent common ancestry of the two species. Findings in this area have been the basis for multiple narratives of human-Neanderthal history that continues to frame comparative studies of both species (Noonan, 2010). The most popular and scientifically accepted theory is that modern human and Neanderthal lineages coexisted and continued to live on evolutionary tracks parallel to one another, prior to their divergence (Noonan, 2010). Descendants of one branch supposedly migrated to Europe and gave rise to Neanderthals, and those of the other branch remained in Africa, eventually becoming modern humans (Noonan, 2010). The modern colonization of Europe, taking place about 40,000 years ago, is said to have potentially brought the two lineages back into widespread contact (Noonan, 2010).

Considering the very recent common ancestry between modern humans and Neanderthals, how much similarity did the two share? Were modern humans and Neanderthals capable of interbreeding? And if so, what outcome resulted? Or is it the case that the species were so distinct that the exchange of meaningful genetic information could not occur? To answer these questions, it is important to discuss the lifestyle of these species. The primary modern humans to colonize Europe had considerably similar cognitive capacities to human beings today (Noonan, 2010). They made cave paintings, figurines, rudimentary musical instruments, and most likely had language (Noonan, 2010). These activities indicate an advanced level of thinking and a highly developed capacity for abstraction (Noonan, 2010). As interesting as this is, such symbolic behaviour was not in fact considered a new phenomenon as some evidence suggests that modern humans were making personal decorations and carving abstract representations in Africa at least 75,000 years ago (Noonan, 2010). Whether Neanderthals had similar abilities and talents is unknown, but they certainly made tools and may have possessed relatively larger brains compared to modern humans (Noonan, 2010). Also, hypotheses that Neanderthals were capable of independently performing these tasks and demonstrating such complex symbolic behaviour are disputed (Noonan, 2010). It is also disputed if Neanderthals actually used language; arguments for and against their language use focus on shared anatomical features shared between humans and Neanderthals (Noonan, 2010).

Limitations of Scientific Techniques

The rudimentary problem in many Neanderthal studies is lack of data, and a reference Neanderthal genome sequence on its own does not allow for the resolution of these questions (Noonan, 2010). What it does provide, when compared with the modern human and chimpanzee reference genomes, is a list of genetic changes unique to modern humans or Neanderthals and changes relative to chimpanzees that are shared between both human species (Noonan, 2010). This comparison helps drive the experimental discovery of biological similarities and differences between modern humans and Neanderthals on a molecular level. The discovery of changes in gene expression or protein function unique to modern humans or Neanderthals serves as an entry point for studying more complex biological phenomena in each lineage, including language and other behavioral traits (Noonan, 2010).

Conclusion

Despite many difficult technical and analytical challenges faced by geneticists, a Neanderthal genome sequence has been generated and has enhanced understanding of the evolutionary relationship between modern humans and

Neanderthals. Multiple scenarios regarding the existence and lifestyle of this species have been presented and continue to be studied today. All in all, the study of human ancestry using recent methods and technologies allows us not only to trace back our ancestry but also to learn and explore the many unique characteristics of modern humans, including their origin and how they came to be part of our DNA. It is important to continue asking questions and educate ourselves on the topic of human genetics, in order to gain insight on, and expand our understanding of our own species.

Chapter 8
Genetic Links between Homo sapiens and Neanderthals
by Omar Abdul Hadi

Introduction

Neanderthals (Homo neanderthalensis) were a group of archaic humans that lived 2.6 million years to 11,700 years ago (Zeberg et al., 2021). Modern humans have Neanderthal DNA in their genome, but the amount of DNA found in modern humans varies by ethnicity (Villanea & Schraiber, 2018). Individuals of East Asian and European descent have the most Neanderthal DNA, where approximately one to two percent of their genome contains Neanderthal DNA (Villanea & Schraiber, 2018). However, individuals of African descent have almost no Neanderthal DNA in their genomes (Villanea & Schraiber, 2018). Ancient Neanderthal genes found in humans today have been shown to lower relative risk of certain viruses, including SARS-CoV-2 (Zeberg et al., 2021).

How Neanderthal Genes were Inherited by Humans

Approximately 20% of Neanderthal genes have survived in the human genome (Villanea & Schraiber, 2018). A single human has an average of 2 to 2.5% Neanderthal DNA (Sankararaman et al., 2016). It is likely that humans have Neanderthal DNA due to interbreeding of the two species from 100,000 years ago (Villanea & Schraiber, 2018). A 2016 study published new Neanderthal gene sequences from the Altai cave in Siberia, as well as from Spain and Croatia (Sankararaman et al., 2016). This set of data shows evidence of human-Neanderthal interbreeding roughly 100,000 years ago. Neanderthal DNA is only found in the nuclear DNA of humans and is not present in the mitochondrial DNA (mtDNA). Mitochondrial DNA is maternally inherited, meaning it can only be passed down from mother to offspring. The mitochondria is a membrane-bound organelle that is present in most eukaryotic cells and is responsible for the production of ATP (Osellame et al., 2012). There are many possible explanations to why this has occurred (Sankararaman et al., 2016). Firstly, there may have been humans that possessed Neanderthal mtDNA at one point in history whose lineage died out. Another possible explanation is that if Neanderthal males were the only ones contributing to the human genome, their genes would not be present in the mtDNA line (Sankararaman et al., 2016). It is also possible that while breeding between Neanderthal males and human females produced fertile offspring, breeding between Neanderthal females and human males did not, meaning that Neanderthal mtDNA could not have been passed down to the next generations (Sankararaman et al., 2016).

Many genes involved in the production of keratin, a protein found in hair, nails, and skin shows high levels of introgression from Neanderthals (Sankararaman et al., 2016). Introgression is the transfer of genetic material from one species to another by the repeated interbreeding between hybrids of the two species (Sankararaman et al., 2016). Many other genes show high levels of introgression from Neanderthals which vary by geographical area. For example, more that 66% of East Asians have a POUF23L variant in their genes while 70% of Europeans have the introgressed allele BNC2 (Zeberg et al., 2020). This chapter explains how certain Neanderthal genes protect against severe symptoms of the SARS-CoV-2 virus, while others can alter the relative risk of developing symptoms in diseases such as type II diabetes, Crohn's disease, biliary cirrhosis and lupus (Zeberg et al., 2020).

Evolutionary History of Neanderthal-Human Interbreeding

It has been hypothesized that the interbreeding of Neanderthals and humans led to both species being exposed to novel viruses and that interbreeding resulted in the exchange of adaptive alleles between the two species that provided protection against these viruses (Enard et al., 2017). Proteins that interact with viruses (virus-interacting proteins or VIPs) tend to adapt at much higher rates than proteins that do not interact with viruses since they are under stronger purifying selection (Enard et al., 2016). Purifying selection (or negative selection) is the selective removal of alleles that are deleterious from the genome. In order to assess whether a haplotype contained an allele from a Neanderthal gene, researchers used a Conditional Random Field (CRF) approach (Enard et al., 2017). CRF is a class of statistical modelling methods often applied in pattern recognition and probability. The objective of the research was to assess the probability that an individual haplotype came from a Neanderthal origin (Enard et al., 2017). Virus-interacting proteins are generally found in areas of the genome where there are high densities of coding sequence and they are usually more highly expressed than non-virus-interacting proteins (Enard et al., 2017). Positive directional selection is a type of natural selection in which the extreme phenotype is selected over other phenotypes. According to the model of positive directional selection, one should observe an increase in virus-interacting proteins in the Neanderthal genome after interbreeding (Enard et al., 2017). The increase in density of virus-interacting proteins in the Neanderthal genome suggests that RNA viruses frequently drove the adaptive introgression between Neanderthals and humans after interbreeding between the two species (Enard et al., 2017). While it is possible that DNA viruses may have caused some introgression of genes between Neaderthals and humans, it is more likely to have been RNA viruses that caused introgression because RNA viruses are more likely to jump between species (Enard et al.., 2017). Researchers then tried to identify which ancient RNA viruses may have caused the introgression between Neanderthals and humans. Human immunodeficiency virus (HIV), Influenza A virus and Hepatitis C virus have the highest number of virus-interacting proteins in the genome; therefore these three ancient RNA viruses may have caused the introgression of certain genes between humans and Neanderthals after interbreeding (Enard et al., 2017).

Link between Neanderthal Genes and the COVID-19 Pandemic

Severe acute respiratory syndrome coronavirus 2 (SARS-CoV-2) is the virus that causes coronavirus disease (COVID-19). Since its discovery in Wuhan, China in late 2019, many individuals who have been infected with COVID-19 have shown varying symptoms and reactions to the novel virus (Zeberg et al., 2020). These symptoms range in severity, stretching from being asymptomatic to having mild symptoms to extreme cases with respiratory failure and death (Zeberg et al., 2020). However, this range of symptoms does not explain why some people experience no symptoms while others experience rapid and extreme sickness. Therefore, there must be genetic factors that play a role in the severity of COVID-19 symptoms. A study conducted in 2020 studied the genetics of 1980 patients that were infected with COVID-19 (Ellinghaus et al., 2020). Scientists identified two genomic regions that are associated with severe COVID-19, one region on chromosome 3 which encodes six genes, and one region on chromosome 9 that determines ABO blood groups (Zeberg et al., 2020). The main gene that is thought to be linked to extreme cases of COVID-19 is SLC6A20. This gene encodes the sodium-imino acid (proline) transporter 1 (SIT1) which interacts with angiotensin-converting enzyme 2 (ACE2). ACE2 is the cellular receptor that the SARS-CoV-2 virus uses to infect cells (Ellinghaus et al., 2020). These six genes encoded on chromosome 3 linked to severe COVID-19 are all in high linkage disequilibrium (LD) (Zeberg et al., 2020). LD refers to the nonrandom association of alleles at different loci (Slatkin et al., 2008). Researchers then attempted to explain whether the Neanderthal haplotype on chromosome 3 was inherited by both Neanderthals and humans due to a common ancestor (Zeberg et al., 2020). The longer a modern human haplotype is shared with Neanderthals, the less likely it is to have been inherited from a common ancestor of the two species. This is because recombination in each generation will break this haplotype into smaller segments (Zeberg et al., 2020). Using statistical analysis, it can be concluded that the risk haplotype (chromosome 3 locus) was not derived from a common ancestor of the two species (Zeberg et al. 2020). Therefore, the chromosome 3 haplotype must have been a direct result of the interbreeding of the two species at some point in history (Zeberg et al., 2020).

A second study concluded that a Neanderthal haplotype located on chromosome 12 is associated with a 22% depletion in relative risk of developing severe COVID-19 symptoms (Zeberg et al., 2021). This haplotype was inherited from Neanderthals at one point in history. The Neanderthal haplotype that protects against severe COVID-19 symptoms is found on chromosome 12 and contains parts of three genes called OAS1, OAS2 and OAS3. This family of genes encode oligoadenylate synthetases (Zeberg et al., 2021). Oligoadenylate synthetases are molecules that regulate early phases of viral infection by destroying viral RNA, resulting in an inhibition of viral infection (Choi et al., 2015). This ancestral Neanderthal locus may have been beneficial throughout Europe and Asia, possibly due to a number of different RNA virus epidemics as this Neanderthal locus has been shown to be protective against three different RNA viruses (West Nile virus, hepatitis C virus, SARS-CoV-2) (Choi et al., 2015). Although these three Neanderthal genes have been preserved, the Neanderthal versions of the OAS gene function differently. A single nucleotide polymorphism (SNP) found in the OAS1 gene affects the splicing of this protein (Zeberg et al., 2021). A SNP is the change of the DNA sequence in one position only. The SNP rs10774671 (location of the SNP) alters splicing of OAS1 transcript such that there are many different isoforms produced rather than the ancestral Neanderthal isoform (Zeber et al., 2021). Neanderthal OAS genes had higher enzymatic activity than the OAS genes commonly present today which may have resulted in a higher protection against various RNA viruses (Zeberg et al., 2021). Today, the Neanderthal haplotype that protects against SARS-CoV-2 has an allele frequency of approximately 30% in Europe and Asia (Zeberg et al., 2021). Genetic analysis shows that this allele frequency has increased significantly very recently. This haplotype had an allele frequency of less than 20% anywhere between 3,000 and 1,000 years ago (Zeberg et al., 2021). Similarly, the Neanderthal chromosome 3 risk locus has also increased in frequency very recently. Genetic analysis shows that prior to 20,000 years ago, there were no carriers of this locus (Zeberg et al., 2021). Today the allele frequency of the Neanderthal chromosome 3 risk locus is approximately 12.5% in individuals of Asian and European descent (Zeberg et al., 2021). This indicates that the Neanderthal OAS locus must have been advantageous to modern humans throughout Europe and Asia (Zeberg et al., 2021). This may have been due to one or many epidemics of RNA viruses in the past. Furthermore, several SNPs on chromosome 12 have been studied and have shown protection against other RNA viruses. The Neanderthal SNP previously discussed (rs10774671) has shown evidence of protection against West Nile virus and increased resistance to Hepatitis C virus (Zeberg et al., 2021). Similarly, a SNP introgressed from the Neanderthal genome on the OAS1 locus (rs2660) has shown to be linked to moderate to high protection against severe COVID-19 symptoms (Zeberg et al., 2021).

Neanderthals were a species of archaic humans that lived in the range of 130,000 to 40,000 years ago. Although Neanderthals no longer exist, approximately 20% of their genetic material is still found in humans today; individual humans on average have 2% Neanderthal DNA in their genome. Neanderthals and humans were genetically similar enough for interbreeding between the two species to occur. As a result of interbreeding, a mixture of their genetic material occurred in a process called genetic introgression. Even after interbreeding, Neanderthal DNA is only present in the nuclear genome of humans and not the mitochondrial genome due to reasons discussed in this chapter. Through natural selection, haplotypes that were beneficial to human survival are still present in the genome today. The introgression of certain genes from Neanderthals have been shown to protect against many modern RNA viruses including West Nile virus, Hepatitis C virus and the novel SARS-CoV-2 virus. During the COVID-19 pandemic, millions of people to date have been infected with the virus. However, everyone experiences different symptoms of the same virus. Therefore there must be a genetic factor that plays a role in the severity of COVID-19 symptoms. Through genetic analysis, researchers discovered a Neanderthal haplotype of chromosome 3 that increases the risk of developing severe COVID-19 symptoms. There were six genes in total that contributed to this increased risk. Conversely, a Neanderthal haplotype on chromosome 12 showed a 22% reduction in the relative risk of developing severe COVID-19 symptoms. This haplotype consists of 3 genes from the same family (OAS1, OAS2, and OAS3). There are many SNPs that contribute to increased protection against COVID-19. Finally, Neanderthal genes remain in the genome because they are beneficial to humans and add a layer of protection against COVID-19 and many other diseases.

Chapter 9
Ethnicity, Neanderthal DNA, and Medicine
by Vedanshi Vala

Introduction

This chapter explores the genetic legacy Neanderthals have left in modern humans, analyzed through the perspectives of ethnicity and medicine. For the purposes of this chapter, anatomically modern humans are distinguished from archaic humans like Neanderthals, through physiological factors such as skull shape and body structure (Race, Ethnicity, and Genetics Working Group, 2005). However, archaic humans—like the Neanderthals—are considered to be biologically and cognitively proximate to modern humans (Johansson, 2014). While our evolutionary ancestors are considered to be recent by comparison to other species, modern humans evolved roughly 200,000 years ago in Africa (Race, Ethnicity, and Genetics Working Group, 2005). Putting things into perspective, this means that anatomically modern humans have existed for a lot of time, even if we have 'recent' archaic ancestors (Neanderthals) by comparison to other mammalian species.

In approaching an investigation so centred on ethnicity, one must first consider the ethics and validity of classifying people into different racial and ethnic groups for research purposes. Given how recent a common ancestor is shared by humans, there is less genetic differentiation amongst modern human groups in different geographic regions than is typically observed in other mammalian species (Race, Ethnicity, and Genetics Working Group, 2005). However, nuances in appearance such as skin colour or hair texture have led to the human construct of 'race' and 'ethnicity', thereby being used to imply the presence of significant genetic differences between different ethnic or racial groups (Race, Ethnicity, and Genetics Working Group, 2005). These categories of race and ethnicity form the foundation of research to ascertain that genetic variation contributes significantly to differences in disease prevalence (Race, Ethnicity, and Genetics Working Group, 2005). It is problematic to rely on these categories solely in such research, which fails to account for the socioeconomic and other such factors that disproportionately affect certain groups, and have a significant impact on health and disease treatment outcomes (Race, Ethnicity, and Genetics Working Group, 2005).

As discussed previously, different racial communities experience differences in susceptibility to disease, and their ability to recover from illness varies based on an assortment of factors going beyond, but also including, genetic makeup. As a result, the medical profession needs to be considerate of such innate factors to the health outcomes of patients based on their ethnicity. There is much that can be discussed on the intersectional topics of ethnicity, Neanderthal DNA, and medicine, and the remainder of this chapter shall endeavour to investigate some such topics of interest.

How Ethnicity Determines the Degree of Archaic Ancestry

When discussing genetics and ancestry, the vessel for human genetic information, DNA, is central to the discussion. DNA is a molecule which carries the genetic information of its organism, its structure having four complementary nucleotide bases, namely: adenine, guanine, cytosine, and thymine (Seeman, 2003). New breakthroughs in chemistry and genetics, and increased understanding of the structural properties of the DNA molecule have increased interest in developing nanotechnology that harnesses its powers (Seeman, 2003).

This exemplifies how DNA could be used by scientists to engineer entirely new possibilities within human genetics in the future, as well as how the inherent properties of DNA make it a crucial database of biological data. As such, the study of DNA yields itself to investigations about the Neanderthal ancestry of modern humans. In particular, a paper discusses findings of Neanderthal genetic information within the chromosome situated in the 3p21.31 region in Eurasians, who are people of both Asian and European descent (Ding et al., 2013). The paper further states that positive selection of the alleles in this region occur with greater frequency in East Asians, with between 49.4% and 66.5% likelihood of positive selection, than in the European population (Ding et al., 2013). On a note of clarity, positive selection refers to instances when certain genetic information is actively expressed. For instance, positive selection for black hair colour and negative selection for red hair colour means the individual will have black hair. Therefore, the previous example demonstrates a higher likelihood to observe active presence of Neanderthal genetic information in East Asians than in Europeans.

As discussed in the chapter introduction, Neanderthals are differentiated from modern humans by a variety of physiological aspects, such as body size; however, they are considered to be genetically similar enough to humans for there to have been interbreeding. There is evidence that East Asian people have a higher degree of Neanderthal ancestry relative to Europeans; that being said, the genetic contribution from Neanderthals is quite small, only being between 1 and 4 percent (Wall et al., 2013). It is also theorized that Neanderthals and modern humans have mixed after the events leading to the ancestral separation of East Asian and European populations (Wall et al., 2013). This mixing may account for the observed differences in genetic ancestry between different ethnic groups (Wall et al., 2013).

In one study, a research group investigated the recency of Neanderthal ancestry in an anatomically modern Romanian human specimen (Fu et al., 2015). The group's publication states that despite Neanderthal disappearance supposedly being 39,000-41,000 years ago, it has been observed that Neanderthal DNA comprises 1-3% of the genome of modern humans in Eurasia (Fu et al., 2015). Their specimen, which itself is dated to be between 37,000-42,000 years old, indicated a presence of 6-9% Neanderthal DNA in its genome (Fu et al., 2015). Moreover, given the size of chromosomal segments of the Neanderthal DNA, the group made the astonishing discovery that this individual may have had a Neanderthal ancestor as recent as four to six generations prior (Fu et al., 2015). That being said, because fossil records of Neanderthals end at the same time that modern humans appear, it remains a mystery as to how these archaic humans were able to mix with their anatomically modern counterparts (Wall et al., 2013).

It is worth noting that there is some controversy regarding the evolutionary passage from archaic to modern humans, with chromosomal evidence indicating one possibility, and fossil evidence telling a different story (Race, Ethnicity, and Genetics Working Group, 2005). It is an oft-heard saying that the mitochondria is the powerhouse of the cell, with mtDNA being mitochondrial DNA, responsible for storing the genetic information powering these organelles (Chen et al., 2010). According to one theory, there are differences in the Y chromosome and parts of the X chromosome, as well as certain autosomal regions of mtDNA, as observed by several researchers (Race, Ethnicity, and Genetics Working Group, 2005). These differences are hypothesized to indicate an absence of significant interbreeding between archaic and modern humans (Race, Ethnicity, and Genetics Working Group, 2005). Conversely, observations of some modern humans sharing physical similarities with archaic humans yields itself to the theory of mixing between Neanderthals and anatomically modern humans (Race, Ethnicity, and Genetics Working Group, 2005). Given these two opposing theories, there is a vacancy of knowledge that further research in this field can be directed towards. Regardless, the presence of Neanderthal DNA in modern humans has several implications, which will be further explored by the following section in this chapter.

Implications of Neanderthal DNA Presence in Anatomically Modern Humans

Findings of archaic DNA from Neanderthals in anatomically modern humans is of academic interest for what it can divulge regarding human genetics, disease susceptibility, and immune response. There is evidence supporting findings that Neanderthal DNA contributes to phenotypic traits, such as skin tone, hair colour, and height as well as behavioural characteristics like sleeping patterns, mood, and even likelihood of smoking habits (Dannemann and Kelso, 2017). Moreover, a study of some 28,000 individuals found that Neanderthal-contributed alleles influence risk factors of several disease symptoms, such as depression, blood-clotting disorders, and lesions of the skin; that being said, this archaic DNA is observed to both exacerbate risks and provide protection against these symptoms (Dannemann and Kelso, 2017). As such, it would be incorrect to think that Neanderthal DNA only increases risk against disease, or that the presence of Neanderthal DNA automatically implies a more robust immune response. There have also been findings that Neanderthal alleles significantly affect height, body fat composition, resting pulse rate, and traits related to skin and hair, with the p value of greater than 1.0×10^{-8} being used as a measure of statistical significance (Dannemann and Kelso, 2017). P values are used by scientists to indicate the probability of getting at least the same results again, and are a means of quantifying significance of data- in other words, a high p value means researchers are very confident in their results. Interestingly, the same group that discovered the aforementioned contributions of Neanderthal DNA also found that there is an extremely small likelihood of the alleles for red hair to have been inherited from archaic ancestors, at just 1.4% Neanderthal allele frequency, whereas the Neanderthal alleles for black hair occur with a whopping 11.0% frequency (Dannemann and Kelso, 2017). This shows that beyond being present in the genetic information of modern humans, these archaic ancestors continue to have influence over variations in the modern human genome with observable phenotypic outputs.

Furthermore, the presence of Neanderthal DNA can potentially have effects on disease susceptibility amongst different racial and ethnic groups. While socioeconomic factors impeding access to healthcare certainly affect health outcomes, genetic makeup can make certain populations more susceptible to cancer than others (Özdemir and Dotto, 2017). This stems from Neanderthals and modern humans in Eurasia interbreeding, resulting in the presence of between 1.5% and 4% of archaic DNA in anatomically modern Eurasians (Özdemir and Dotto, 2017). The composition of the genome of African Americans results in a higher susceptibility to cancer due to genetic predispositions to disease like obesity, as well as to chronic inflammation (Özdemir and Dotto, 2017). DNA acquired from Neanderthal ancestors may further be a cause for an overactive immune system in African people, which leads to greater autoimmune disease susceptibility relative to Europeans (Reardon, 2016). This point in particular has been explored further in academia during the COVID-19 pandemic, stemming from observed differences in mortality from the virus between European and East Asian ethnic groups (Yamamoto et al., 2021). On the other hand, a particular active Neanderthal allele in the European population makes them more prone to sunburns, which is a risk factor for skin cancer (Dannemann and Kelso, 2017). Moreover, a study hypothesized that modern humans developed their own unique immune response upon encountering different pathogens in Europe (Reardon, 2016). This process of adapting to the different environment in Europe would have been made easier with the immune response obtained from Neanderthals, through a process termed as 'immune mixing' (Reardon, 2016). Immune mixing in this sense refers to Europeans passing Neanderthal immune responses to their descendants as a result of interbreeding (Reardon, 2016). There have further been observations of Europeans exhibiting a similar immune response as Neanderthals, whereas the immune response of Africans is different, speaking to the contrasting situations in which these discrepancies in immune responses were developed (Reardon, 2016).

As such, rather than one specific ethnic or racial group being predisposed to all types of illness and disease, all ethnic groups have varying susceptibility to different diseases. Some diseases may be more easily contracted by people of African descent, while others may affect Asian populations more severely, and others still may be extremely dangerous to European groups. There are, as such, varying considerations that must be accounted for in healthcare, which will be further explored in the next section.

Genetics and Ethnicity: Considerations for Medical Education, Practice, and Treatment

The systemic issues present in healthcare start with the inequity and discrimination experienced first by healthcare professionals, instructors, and students before they reach patients. In a study with 31 subjects, all of whom were black medical students, 30 of them reported experiencing racism at some point in their medical education, highlighting the severity of racism in higher education (Bullock and Houston, 1987). In another study examining discrimination against practicing physicians, 63% of participating physicians reported workplace discrimination (Coombs and King, 2005). Of all respondents to the survey, 42% of physicians were from racial or ethnic minorities (Coombs and King, 2005). As such, the system which enables such injustice must first be recalibrated. Such changes would not only create a more inclusive, respectful, and welcoming environment for practicing physicians and medical students, but consequently lead to a more positive patient experience as well, because the entirety of the medical system is connected.

The importance of transforming healthcare and pre-healthcare education in order to achieve health equity has been impounded by several publications in academia. For one, it has been ideated that the forgoing of certain topics, such as that of social factors affecting health, from the curriculum taught in schools for health professionals "conveys an implicit message about the health and health outcomes of people of colour" (Ackerman-Barger et al., 2020, p. 1).

The same authors propose that this so-termed 'hidden curriculum' be taught to students in health programs to break the stigma around these topics; specifically, the paper states that medical professionals having a better understanding of the correlation between social and structural aspects to disease can pave the road to better health outcomes for marginalized and oppressed groups (Ackerman-Barger et al., 2020). Another paper suggests that an approach to tackling this change in curriculum effectively is by implementing structural competency training in undergraduate programs, so that students are already prepared to enter the healthcare field with a certain mindset (Metzl et al., 2018). Structural competency training enables participants to consider the various socioeconomic, or 'structural', factors that influence health outcomes, such as systemic racism and the health inequities it exacerbates (Metzl et al., 2018). Diversifying the racial and ethnic composition within the medical field is also crucial to reducing implicit bias from physicians, which can contribute to decisions being made based on cultural stereotypes rather than individual patient requirements (Chapman et al., 2013). It should be the responsibility of the educational system to equip future physicians and medical professionals with the ability to recognize their implicit biases, and the tools to make informed and case-dependent judgments for all patients.

From the patient's perspective, ethnicity can be a significant determinant of their health outcomes, and as such, must be given adequate weightage in the medical diagnosis and treatment process. For one, the different genetic makeup of all ethnic groups can alter their tolerance and response to certain medication (Neary and Owen, 2017). An example of such a situation would be in obtaining higher success rates of antiretroviral therapy for HIV treatment, where different ethnicities may respond in varying manners to the key compounds in the medication (Neary and Owen, 2017). In particular, a study investigated the effects that enzymes contained in antiretroviral drugs have in various ethnic groups, with findings increasing understanding about how genetic variation affects an individual's immune response to these drugs (Neary and Owen, 2017). The more tailored antiretroviral therapy is to the patient's genome, the higher the chance of their body having a long-lasting immune response against HIV, which further iterates the importance of ethnicity being factored into medical treatment plans (Neary and Owen, 2017). There have, moreover, been findings evidencing that different ethnic groups experience pain differently (Edwards et al., 2001). For instance, in a comparison between African-Americans and non-Hispanic Caucasians, a study found that African-Americans had a lower tolerance for thermal stimuli (Edwards et al., 2001). This finding is of significance because it contradicts the previously-held racist belief regarding lower pain perception in black people, which has historically resulted in biased decision-making for pain treatment in black people (Hoffman et al., 2016). Understanding these differences in pain thresholds can aid physicians in their approach to treatment options, as well as in their understanding that individuals of certain ethnicities may find a certain event or incident to be extremely painful, whereas it would be experienced entirely differently by another ethnic group. Moreover, an ethnic group's ability to access support services, such as those for mental health, may be challenged due to stigmas associated with mental illness and mental health care services (Neighbors et al., 2007). In a study, it was found that only 10.1% of participants, who consisted of black people living in the United States, used mental health services in the year prior (Neighbors et al., 2007). The paper in which this study was published further divulged how people with insurance are more likely to use certain services than those without it, which highlights how economic factors, which disproportionately affect certain ethnic groups, can be a barrier to accessing medical care (Neighbors et al., 2007). Yet another aspect to medical care is to provide solutions that align with the patient's personal beliefs. When difficult decisions surrounding life-sustaining technology need to be made, it is made more difficult if there are significant cultural differences between the doctor and patient, with communication being impeded due to a discrepancy in values (Blackhall et al., 1999). As demonstrated by these examples, medical care must be provided in an ethnically sensitive manner which accounts for differences in genetics, barriers to accessibility, and cultural belief systems.

All in all, ethnicity has a high degree of influence over the health outcomes of patients, as well as the experiences of medical students and practitioners in their educational institutions and places of work. As such, ethnicity and its implications on the entirety of the medical system must be continually addressed to provide the optimal learning, practitioning, and healthcare experience for all those involved.

Conclusion

This chapter explored ethnicity, Neanderthal DNA, and medicine, focusing on a few key topics of interest. Firstly, the biological differentiation between anatomically modern humans and their archaic ancestors was made through certain anatomical features. Moreover, in introducing the importance of approaching this conversation ethically, race and ethnicity were classified as a human construct, having limited implications on genetic differentiation amongst anatomically modern humans. Then, the first section of this chapter discussed the connection between ethnicity and inheritance of archaic DNA from Neanderthals, followed by which implications of the presence of such genetic information were expanded upon. To reiterate, Neanderthal DNA presence in anatomically modern humans impacts certain phenotypic consequences, such as physical appearance, as well as disease predisposition and immune response amongst different ethnic groups. The final section of this chapter discussed, in light of the information that ethnic groups are affected by illnesses differently, medical care being made more equitable to achieve better healthcare outcomes for people of racial minorities.

In closing, this chapter explored some of the points of interest in the intersecting discussion of race, inheritance of archaic genetic information, and equitable healthcare. As previously discussed in this chapter, there is much work that needs to be done to build communities that can better cater to the unique needs of all people in them. Understanding the evolution of anatomically modern humans from our Neanderthal ancestors, and the implications this has on our genome, as well as the ugly reality of inequity and discrimination in the medical system, is one of the first steps towards such an ambition. It is also worth iterating that understanding genetic differences can pave the way to more customized healthcare; however, as stated at the very start of this chapter, race and ethnicity is ultimately a human construct based on cultural differences and outward appearances. Genetic differences account for only a fraction of what it is to be human. Race and ethnicity is but one facet of the human story. Human beings are, in reality, closer to each other than it seems on the surface, and as one humankind, we should strive towards an equitable world for all.

Chapter 10
The Neanderthals and Potential Causes of their Extinction
by Joylen Kingsley

Introduction

Neanderthals were closely related to humans but contain enough genetic and phenotypic differences to be classified as a separate species. Neanderthals were first discovered in 1856: they lived approximately 130 000 to 40 000 years ago, and their relatively recent existence compared to alternative human like species, allows a richer exploration into the history of the Homo neanderthalensis species (Trinkaus & Howells, 1979). One aspect of our understanding regarding Neanderthals is the lack of definitive reasoning behind their extinction. There are many competing hypotheses, but no conclusions have been acknowledged as the true cause of Neanderthal extinction (Timmermann, 2020). Fields from archaeology to pathology and even economics have theories regarding the extinction of Neanderthals (Timmermann, 2020).

This chapter in particular explores the evidence behind the various theories contributing to Neanderthal extinction, including: the competitive exclusion theory, assimilation, demographic weakness, climate change and pathogens (Timmermann, 2020). While the purpose of this chapter is not to emphasize one theory over another, presenting these hypotheses adjacently will allow for a deeper understanding of the potentially collaborative nature of said hypotheses. Researchers base their conclusions on models, heavy with educated assumptions; therefore the presented theories must be considered with caution.

Competitive Exclusion as a Cause of Neanderthal Extinction

Despite anatomical similarities and ongoing debates, Homo sapiens, or anatomically modern humans (AMH), are a different species from Homo neanderthalensis (Banks et al., 2008). The similarities between both modern humans and the Neanderthal species included the regions they inhabited and the resources they employed, especially in the Eurasian geographic region (Gilpin et al., 2016). When two similar species that compete for the same resources live in close proximity, it usually leads to interspecific competition, or competition between species for limited resources (Gilpin et al., 2016). This principle can be further demonstrated by the Lotka-Volterra model of interspecific competition, where one species lives in relative success while the outcompeted species is eradicated, sometimes even to extinction, as was the theory with the Neanderthals (Gilpin et al., 2016). The competitive exclusion theory suggests that the advanced tools and techniques employed by anatomically modern humans allowed them to outcompete the Neanderthals (Goldfield et al., 2018).

Initially, anatomically modern humans were the smaller population, but they expanded their reach through iterative propagule, a process in which small groups of people break off and propagate their culture and practices in various post-migratory settlements (Gilpin et al., 2016). This model allowed anatomically modern humans to increase genetic variety, which is further supported by the increased heterozygosity as populations veer further away from Africa (Gilpin et al., 2016). Heterozygosity refers to an increased number of possible genetic combinations resulting in both visible and invisible changes in anatomically modern humans. The competitive exclusion theory suggests that anatomically modern humans were initially lesser in population, but a combination of genetic variation, migration and superior use of tools allowed them the advantage over Neanderthals (Banks et al., 2008; Gilpin et al., 2016).

One of the defining features of human survival was the use of fire (Goldfield et al., 2018). Unfortunately, the use of fire in the Middle Paleolithic era is limited (Goldfield et al., 2018). Research indicates that only a small percentage of settlements among the Neanderthals used fire and even then, fire was usually started by natural causes, such as lightning strikes (Goldfield et al., 2018). Even if Neanderthals had been able to maintain a naturally started fire, they lived primarily in a snowy, cold climate, preventing the prolonged maintenance of fire (Goldfield et al., 2018). Records of fire use and research into the area is scarce in the Middle Paleolithic era, but regardless of Neanderthal proficiency with fire, anatomically modern humans most definitely had the ability to start fires (Goldfield et al., 2018). This ability in and of itself served as a competitive advantage between species. While fire serves a number of survival benefits, one of the major innovations of anatomically modern humans was their ability to cook. Digesting raw foods is biologically and energetically taxing (Goldfield et al., 2018). Cooking externalizes part of the digestion process, allowing greater energy conservation. This conservation also served as an evolutionary advantage over Neanderthals, who lacked the ability to harness fire for alternative uses, including cooking (Goldfield et al., 2018).

Assimilation and Demographic Weakness as a Cause of Neanderthal Extinction

Alternatively, the assimilation model is another possible cause of Neanderthal extinction (Smith et al., 2005). The model was first introduced in the 1980s and suggests that during their migration from Africa to Asia and Europe, anatomically modern humans interacted with the Neanderthals, resulting in an exchange of genetic material (Smith et al., 2005). So much so, that modern features in anatomically modern humans may be a result of intermingling with late age Neanderthals (Smith et al., 2005). As more is discovered about the Neanderthals, opinions continue to change, and opposing theories continue to exist, but assimilation is an alternative to competitive exclusion. It must be remembered that there were significantly less Neanderthals than there were anatomically modern humans; therefore, mating would have been a reasonable course of action towards genetic integration (Smith et al., 2005). Fertile ages for Neanderthal women ranged from aged 18 to 30, with the 18- to 20-year-old group being slightly less likely to reproduce than the 21- to 30-year-old group (Degioanni et al., 2019). Slight reduction in average fertility rate resulted in lowered growth rate, and during the middle Paleolithic period, many factors were in constant motion, resulting in fluctuating fertility (Degioanni et al., 2019). These factors could include nutrient stress due to cold temperatures, and minimal mating choices due to internal conflict and hunting incidents (Degioanni et al., 2019). While assimilation is only a possible contributor to extinction, demographic weakness must also be considered as a driving force towards Neanderthal extinction (Degioanni et al., 2019).

Climate Change and Vegetation Change as a Cause of Neanderthal Extinction

It is easy to cite the intrinsic nature of human-like species for features such as competition and the urge to maintain population through mating, but those urges were not the sole reason behind Neanderthal extinction. Climate change and vegetation changes are another theory supporting the eradication of Neanderthals in the middle Paleolithic period (Finlayson & Carrión, 2007). The last glaciation 50 000 to 12 000 years ago coincides with the extinction of Neanderthals and the rise of anatomically modern humans (Finlayson & Carrión, 2007). The Palearctic ecozone, ranges from modern North Africa to Modern North Europe, and 50 000 years ago only Neanderthals inhabited these areas (Finlayson & Carrión, 2007) Around 30 000 years ago, a majority of the same region was primarily inhabited by anatomically modern humans, with only small congregations of Neanderthals remaining (Finlayson & Carrión, 2007). Climate change in the middle Paleolithic era was fast and unpredictable, due to fluctuations in pollen levels that further affected vegetation, and animal life that depended on vegetation (Finlayson & Carrión, 2007). The southern regions of the Palearctic ecozone were known for sections of humid climate with wood-based vegetation opposingly, the glacial climate tended to dominate more in the Northern and Eastern parts of Europe (Finlayson & Carrión, 2007). The temperature paired with the forest population in the Southern region led to increased Neanderthal settlements and animal populations (Finlayson & Carrión, 2007).

The climate of the middle Paleolithic period was characterized by cycles of glaciation. Neanderthals were cold adapted, but only to milder weather, as opposed to the severe cold (Gilligan, 2007). Researchers believe that despite their lack of knowledge with tools, Neanderthals were able to create simple clothing, classified as one or two layered clothing typically draped over one's body (Gilligan, 2007). Any warmer clothing would have resulted in potentially differing evolution, in regards to the Neanderthals cold adaptation (Gilligan, 2007). Although this should have helped the Neanderthals sufficiently, the time period hosted many intense temperature fluctuations resulting in cold stress on the Neanderthal bodies (Gilligan, 2007). Unfortunately, hypothermia does not leave obvious pathological signs on the Neanderthal body; similarly, resistance to frostbite was one of the Neanderthal's evolutionary advantages due to their cold adaptation (Gilligan, 2007). Therefore, from an archeological perspective with respect to bone analysis, the extreme cold cannot definitively be cited as a cause for Neanderthal extinction (Gilligan, 2007). Although, with the exception of the glacial cycles, there was little need to create thicker clothing, therefore the sudden fluctuations of cold temperatures, and lack of preparation, may have played a significant role in extinction, reinforcing climate change as a credible theory (Gilligan, 2007).

Pathogens and Transmissible Spongiform Encephalopathy as a Cause of Neanderthal Extinction

Pandemics through time have limited the population, with the most recent example being COVID-19. Even in modern times, humans are plagued by disease, the Neanderthals were no exception to this rule. Neanderthals were very strong hunters, making it very likely that their diet primarily consisted of meat (Chiarelli, 2004). It is less commonly known that Neanderthals actually practiced cannibalism, potentially for ritualistic purposes (Chiarelli, 2004). Neanderthals were also known to consume the brains of many of their prey, including deer. This paired with their minimal ability with fire implies heavy consumption of raw meat (Chiarelli, 2004). These facts could indicate transmissible spongiform encephalopathy, commonly seen in cows, but ultimately a neurological disorder carried through prions (Chiarelli, 2004). Prions are proteins that can be transferred between species, and can cause nerve cell degeneration (Chiarelli, 2004). Cannibalism among Neanderthals dates back to as far as 47 thousand years ago, although the exact rate of the practice among the population is unknown (Timmermann, 2020).

Transmissible spongiform encephalopathy was first recorded in the 1900s, in Papua New Guinea, among a hunter gatherer group, with similar cannibalistic practices to the Neanderthals (Underdown, 2008). Transmissible spongiform encephalopathy has a relatively long incubation period, therefore it is difficult to immediately connect cannibalism to the disease (Underdown, 2008). 'Kuru' is another name for transmissible spongiform encephalopathy and as such is the name of the model used to extrapolate the effect of the disease among Neanderthals (Underdown, 2008). Population dynamics depend on the viable sample among the total population, 15 000 transmissible spongiform encephalopathy related deaths could reduce reproductive populations within 250 years (Underdown, 2008). Transmissible spongiform encephalopathies are deadly among those that are infected, but the disease can exist in populations and sporadically reoccur, further affecting population growth (Underdown, 2008). It is impossible to determine the reason behind cannibalism within Neanderthal populations, but there are two prevalent theories; the 'competition' perspective and the 'need driven' perspective (Underdown, 2008). The 'competition' theory of cannibalism may have been ritualistic, done to develop a hierarchy among hunter gatherer groups, alternatively the 'need driven' theory of cannibalism may have been practiced for the purpose of nourishment when resources were scarce (Underdown, 2008). The 'need driven' theory suggests a more rapid form of infection among a population, but regardless both theories provide equal opportunity for spread (Underdown, 2008). Lack of hygienic practices and consistent opportunities for infection would have eventually affected Neanderthal population (Underdown, 2008).

Ultimately, as hunter gatherers, Neanderthals depended heavily on managing population numbers to maintain the demands of the population. Researchers cannot prove transmissible spongiform encephalopathy as the cause of extinction, but at the very least the disease may have contributed to weakening the viable population (Underdown, 2008).

Concluding Remarks Regarding The Neanderthal Extinction

As new research comes to light, various theories are proved and disproved regarding Neanderthal extinction. It may even be accurate to suggest that rather than one distinct extinction event, multiple factors may have contributed to Neanderthal extinction. Many conclusions have been drawn regarding competition, climate and pathogens, but majority of these theories and their proofs stem from models. It is impossible to truly understand Neanderthal life without having lived in that time period, but opposing theories can be threaded to create a relatively clear picture.

The most common theories regarding Neanderthal extinction are the claims of competitive exclusion by Homo sapiens, assimilation, demographic weakness, climate change, vegetation change, and pathologies. Researchers understand Neanderthals because of their relative youth in terms of history. We must learn about the Neanderthals to better understand our own origins as humans. While we do not yet have enough information regarding the exact cause of their extinction, that may be subject to change as we discover more.

References

Chapter 1

Durmaz, A. A., Karaca, E., Demkow, U., Toruner, G., Schoumans, J., Cogulu, O. (2015). Evolution of genetic techniques: past, present, and beyond. Biomed Res Int. 2015;2015(461524). doi:10.1155/2015/461524

Venter, J.C., Adams, M. D., Myers, E. W., Li, P. W., mural, R. J., Sutton, G. G., Smith, H. O., Yandell, M., Evans, C. A., Holt, R. A., Goocayne, J. D., Amanatides, P., Ballew, R. M., Huson, D. H., Wortman, J. R., Zhhang, Q., Kodira, C. D. Zheng, X. H., Chen, L., Skupski, M., ...Zhu, X. (2001). The Sequence of the Human Genome. Science, 291(5507), 1304-1351. 10.1126/science.1058040

Chapter 2

Gee, H. (2000). Neanderthal DNA confirms distinct history. Nature. https://doi.org/10.1038/news000030-8

Hublin, J. (2009). The origin of Neandertals. Proceedings Of The National Academy Of Sciences, 106(38), 16022-16023. https://doi.org/10.1073/pnas.0904119106

Madison, P. (2016). The most brutal of human skulls: Measuring and knowing the first Neanderthal. The British Journal For The History Of Science, 49(3), 411-430. https://doi.org/10.1017/s0007087416000650

Mellars, P. (1998). The fate of the Neanderthals. Nature, 395(6702), 539-540. https://doi.org/10.1038/26842

Rogers, A., Bohlender, R., & Huff, C. (2017). Early history of Neanderthals and Denisovans. Proceedings Of The National Academy Of Sciences, 114(37), 9859. https://doi.org/10.1073/pnas.1706426114

Soressi, M. (2016). Neanderthals built underground. Nature, 534(7605), 43-44. https://doi.org/10.1038/nature18440

Tattersall, I., & Schwartz, J. (1999). Hominids and hybrids: The place of Neanderthals in human evolution [Peer commentary on the article "The early Upper Paleolithic human skeleton from the Abrigo do Lagar Velho (Portugal) and modern human emergence in Iberia" by C. Duarte et al.]. https://www.pnas.org/content/96/13/7117#ref-8.

Weyrich, L., Duchene, S., Soubrier, J., Arriola, L., Llamas, B., & Breen, J. et al. (2017). Neanderthal behaviour, diet, and disease inferred from ancient DNA in dental calculus. Nature, 544(7650), 357-361. https://doi.org/10.1038/nature21674

Chapter 3

Appenzeller, T. (2013). Neanderthal culture: Old masters. Nature, 497(7449), 302–304. https://doi.org/10.1038/497302a

Douka, K., & Spinapolice, E. E. (2012). Neanderthal Shell Tool Production: Evidence from Middle PALAEOLITHIC Italy and Greece. Journal of World Prehistory, 25(2), 45–79. https://doi.org/10.1007/s10963-012-9056-z

Ledford, H. (2007). Some Neanderthals were red-heads. Nature. https://doi.org/10.1038/news.2007.197

Marín, J., Saladié, P., Rodríguez-Hidalgo, A., & Carbonell, E. (2017). Neanderthal hunting strategies inferred from MORTALITY profiles within THE Abric Romaní sequence. PLOS ONE, 12(11). https://doi.org/10.1371/journal.pone.0186970

Mellars, P. (2004). Neanderthals and the modern human colonization of Europe. Nature, 432(7016), 461–465. https://doi.org/10.1038/nature03103
Rodríguez-Hidalgo, A., Morales, J. I., Cebrià, A., Courtenay, L. A., Fernández-Marchena, J. L., García-Argudo, G., Marín, J., Saladié, P., Soto, M., Tejero, J.-M., & Fullola, J.-M. (2019). The châtelperronian Neanderthals of COVA Foradada (Calafell, Spain) used imperial Eagle phalanges for symbolic purposes. Science Advances, 5(11). https://doi.org/10.1126/sciadv.aax1984

Weyrich, L. S., Duchene, S., Soubrier, J., Arriola, L., Llamas, B., Breen, J., Morris, A. G., Alt, K. W., Caramelli, D., Dresely, V., Farrell, M., Farrer, A. G., Francken, M., Gully, N., Haak, W., Hardy, K., Harvati, K., Held, P., Holmes, E. C., … Cooper, A. (2017). Neanderthal behaviour, diet, and disease inferred from ancient DNA in dental calculus. Nature, 544(7650), 357–361. https://doi.org/10.1038/nature21674

Yotova, V., Lefebvre, J.-F., Moreau, C., Gbeha, E., Hovhannesyan, K., Bourgeois, S., Bedarida, S., Azevedo, L., Amorim, A., Sarkisian, T., Avogbe, P. H., Chabi, N., Dicko, M. H., Kou' Santa Amouzou, E. S., Sanni, A., Roberts-Thomson, J., Boettcher, B., Scott, R. J., & Labuda, D. (2011). An X-Linked Haplotype of Neandertal origin is present among All non-african populations.

Molecular Biology and Evolution, 28(7), 1957–1962. https://doi.org/10.1093/molbev/msr024

Chapter 4

Benjafield, J. G., Smilek, D., & Kingstone, A. (2010). Cognition. Oxford University Press.

Caron, F., d'Errico, F., Del Moral, P., Santos, F., & Zilhão, J. (2011). The reality of NEANDERTAL SYMBOLIC behavior at THE Grotte du Renne, Arcy-sur-Cure, France. PLoS ONE, 6(6). https://doi.org/10.1371/journal.pone.0021545

Churchill, S. E. (2006). Bioenergetic perspectives on Neanderthal THERMOREGULATORY and ACTIVITY BUDGETS. Neanderthals Revisited: New Approaches and Perspectives, 113–133. https://doi.org/10.1007/978-1-4020-5121-0_7

Cohen, M. D., & Axelrod, R. (1984). Coping with Complexity: The Adaptive Value of Changing Utility. The American Economic Review, 74(1), 30–42.

Henshilwood, C. S., & Marean, C. W. (2003). The origin of modern human behavior. Current Anthropology, 44(5), 627–651. https://doi.org/10.1086/377665

Horan, R. D., Bulte, E., & Shogren, J. F. (2005). How trade saved humanity from biological Exclusion: An economic theory of Neanderthal extinction. Journal of Economic Behavior & Organization, 58(1), 1–29. https://doi.org/10.1016/j.jebo.2004.03.009

Izard, C. E., B., K. J. Z. R., & Lang, P. J. (1990). Cognition in emotion: concept and action. In Emotions, cognition and behavior. essay, Cambridge University Press.

Mellars, P. (2005). The impossible coincidence. A Single-species model for the origins of modern human behavior in Europe. Evolutionary Anthropology: Issues, News, and Reviews, 14(1), 12–27. https://doi.org/10.1002/evan.20037

MORO ABADÍA, O. S. C. A. R., & GONZÁLEZ MORALES, M. A. N. U. E. L. R. (2010). Redefining neanderthals and art: An alternative interpretation of the multiple species model for the origin of behavioural modernity. Oxford Journal of Archaeology, 29(3), 229–243. https://doi.org/10.1111/j.1468-0092.2010.00346.x

Rendu, W., Renou, S., Soulier, M.-C., Rigaud, S., Roussel, M., & Soressi, M. (2019). Subsistence strategy changes during the middle to Upper Paleolithic Transition reveals specific adaptations of human populations to their environment. Scientific Reports, 9(1). https://doi.org/10.1038/s41598-019-50647-6

Chapter 5

Bailey, S. E. (2002). A closer look at Neanderthal postcanine dental morphology: The mandibular dentition. The Anatomical Record, 269(3), 148–156. doi:10.1002/ar.10116

Clement, A. F., Hillson, S. W., & Aiello, L. C. (2012). Tooth wear, Neanderthal facial morphology and the anterior dental loading hypothesis. Journal of Human Evolution, 62(3), 367–376. doi:10.1016/j.jhevol.2011.11.014

Demes, B. (1987). Another look at an old face: biomechanics of the neandertal facial skeleton reconsidered. Journal of Human Evolution, 16(3), 297–303. doi:10.1016/0047-2484(87)90005-4

Franciscus, R. G. (2003). Internal nasal floor configuration in Homo with special reference to the evolution of Neandertal facial form. Journal of Human Evolution, 44(6), 701–729. doi:10.1016/s0047-2484(03)00062-9

Frayer, D. W., & Russell, M. D. (1987). Artificial grooves on the Krapina Neanderthal teeth. American Journal of Physical Anthropology, 74(3), 393–405. doi:10.1002/ajpa.1330740311

Holton, N. E., & Franciscus, R. G. (2008). The paradox of a wide nasal aperture in cold-adapted Neandertals: a causal assessment. Journal of Human Evolution, 55(6), 942–951. doi:10.1016/j.jhevol.2008.07.001

Krueger, K. L., Willman, J. C., Matthews, G. J., Hublin, J.-J., & Pérez-Pérez, A. (2019). Anterior tooth-use behaviors among early modern humans and Neandertals. PLOS ONE, 14(11), e0224573. doi:10.1371/journal.pone.0224573

O'Connor, C. F., Franciscus, R. G., & Holton, N. E. (2005). Bite force production capability and efficiency in Neandertals and modern humans. American Journal of Physical Anthropology, 127(2), 129–151. doi:10.1002/ajpa.20025

Relethford, J.H. (2008). Genetic evidence and the modern human origins debate. Heredity, 100, 555–563. https://doi.org/10.1038/hdy.2008.14

Spencer, M. A., & Demes, B. (1993). Biomechanical analysis of masticatory system configuration in Neanderthals and Inuits. American Journal of Physical Anthropology, 91(1), 1–20. doi:10.1002/ajpa.1330910102

Weaver, T. D. (2009). The meaning of Neandertal skeletal morphology. Proceedings of the National Academy of Sciences, 106(38), 16028–16033. doi:10.1073/pnas.0903864106

Chapter 6

Bamshad, M., & Wooding, S. P. (2003). Signatures of natural selection in the human genome. Nature Reviews Genetics, 4(2), 99–110. https://doi.org/10.1038/nrg999

Biswas, S., & Akey, J. M. (2006). Genomic insights into positive selection. Trends in Genetics, 22(8), 437–446. https://doi.org/10.1016/j.tig.2006.06.005
Joly, Y., Ngueng Feze, I., & Simard, J. (2013). Genetic discrimination and life insurance: A systematic review of the evidence. BMC Medicine, 11(1). https://doi.org/10.1186/1741-7015-11-25

Vitti, J. J., Cho, M. K., Tishkoff, S. A., & Sabeti, P. C. (2012). Human evolutionary genomics: Ethical and interpretive issues. Trends in Genetics, 28(3),137–145. https://doi.org/10.1016/j.tig.2011.12.001
Spencer H. Principles of Biology. 1865;1:530–531.

Yi, X., Liang, Y., Huerta-Sanchez, E., Jin, X., Cuo, Z. X., Pool, J. E., Xu, X., Jiang, H., Vinckenbosch, N., Korneliussen, T. S., Zheng, H., Liu, T., He, W., Li, K., Luo, R., Nie, X., Wu, H., Zhao, M., Cao, H., … Wang, J. (2010). Sequencing of 50 Human exomes reveals adaptation to high altitude. Science, 329(5987), 75–78. https://doi.org/10.1126/science.1190371

Chapter 7

Herrera, K. J., Somarelli, J. A., Lowery, R. K., & Herrera, R. J. (2009). To what extent did Neanderthals and modern Humans interact? Biological Reviews, 84(2), 245–257. https://doi.org/10.1111/j.1469-185x.2008.00071.x

Noonan, J. P. (2010). Neanderthal genomics and the evolution of modern humans. Genome Research, 20(5), 547–553. https://doi.org/10.1101/gr.076000.108

Noonan, J. P., Coop, G., Kudaravalli, S., Smith, D., Krause, J., Alessi, J., Chen, F., Platt, D., Paabo, S., Pritchard, J. K., & Rubin, E. M. (2006). Sequencing and analysis of neanderthal genomic dna. Science, 314(5802), 1113–1118. https://doi.org/10.1126/science.1131412

Sankararaman, S., Mallick, S., Dannemann, M., Prüfer, K., Kelso, J., Pääbo, S., Patterson, N., & Reich, D. (2014). The genomic landscape of Neanderthal ancestry in present-day humans. Nature, 507(7492), 354–357. https://doi.org/10.1038/nature12961

Chapter 8

Choi, U. Y., Kang, J.-S., Hwang, Y. S., & Kim, Y.-J. (2015). Oligoadenylate synthase-like (OASL) proteins: dual functions and associations with diseases. Experimental & Molecular Medicine, 47(3). https://doi.org/10.1038/emm.2014.110

Ellinghaus, D. (2020). Genomewide Association Study of Severe Covid-19 with Respiratory Failure. New England Journal of Medicine, 383(16), 1522–1534. https://doi.org/10.1056/nejmoa2020283

Enard, D., Cai, L., Gwennap, C., & Petrov, D. A. (2016). Viruses are a dominant driver of protein adaptation in mammals. ELife, 5. https://doi.org/10.7554/elife.12469

Enard, D., & Petrov, D. A. (2017). RNA viruses drove adaptive INTROGRESSIONS between Neanderthals and modern humans. https://doi.org/10.1101/120477

Osellame, L. D., Blacker, T. S., & Duchen, M. R. (2012). Cellular and molecular mechanisms of mitochondrial function. Best Practice & Research Clinical Endocrinology & Metabolism, 26(6), 711–723. https://doi.org/10.1016/j.beem.2012.05.003

Sankararaman, S., Mallick, S., Patterson, N., & Reich, D. (2016). The combined landscape of Denisovan and Neanderthal ancestry in Present-Day Humans. Current Biology, 26(9), 1241–1247. https://doi.org/10.1016/j.cub.2016.03.037

Slatkin, M. (2008). Linkage disequilibrium — understanding the evolutionary past and mapping the medical future. Nature Reviews Genetics, 9(6), 477–485. https://doi.org/10.1038/nrg2361

Villanea, F. A., & Schraiber, J. G. (2018). Multiple episodes of interbreeding between Neanderthal and modern humans. Nature Ecology & Evolution, 3(1), 39–44. https://doi.org/10.1038/s41559-018-0735-8

Zeberg, H., & Pääbo, S. (2020). The major genetic risk factor for severe COVID-19 is inherited from Neanderthals. Nature, 587(7835), 610–612. https://doi.org/10.1038/s41586-020-2818-3

Zeberg, H., & Pääbo , S. (2021). A genomic region associated with protection against severe COVID-19 is inherited from Neandertals. Proceedings of the National Academy of Sciences. https://doi.org/e2026309118

Chapter 9

Ackerman-Barger, K., London, M., & White, D. J. (2020). When an omitted curriculum becomes a hidden curriculum: Let's teach to promote health equity. Journal of Health Care for the Poor and Underserved, 31(4S), 182–192. https://doi.org/10.1353/hpu.2020.0149

Blackhall, L. J., Frank, G., Murphy, S. T., Michel, V., Palmer, J. M., & Azen, S. P. (1999). Ethnicity and attitudes towards life sustaining technology. Social Science & Medicine, 48(12), 1779–1789. https://doi.org/10.1016/s0277-9536(99)00077-5

Bullock, S. C., & Houston, E. (1987). Perceptions of racism by black medical students attending white medical schools. Journal of the National Medical Association, 79(6), 601–608.

Chapman, E. N., Kaatz, A., & Carnes, M. (2013). Physicians and implicit bias: How doctors may unwittingly perpetuate health care disparities. Journal of General Internal Medicine, 28(11), 1504–1510. https://doi.org/10.1007/s11606-013-2441-1

Chen, H., Vermulst, M., Wang, Y. E., Chomyn, A., Prolla, T. A., McCaffery, J. M., & Chan, D. C. (2010). Mitochondrial fusion is required for mtDNA stability in skeletal muscle and tolerance of mtDNA mutations. Cell, 141(2), 280–289. https://doi.org/10.1016/j.cell.2010.02.026

Coombs, A. A., & King, R. K. (2005). Workplace discrimination: experiences of practicing physicians. Journal of the National Medical Association, 97(4), 467–477.

Dannemann, M., & Kelso, J. (2017). The contribution of Neanderthals to phenotypic variation in modern humans. The American Journal of Human Genetics, 101(4), 578–589. https://doi.org/10.1016/j.ajhg.2017.09.010

Ding, Q., Hu, Y., Xu, S., Wang, J., & Jin, L. (2013). Neanderthal introgression at Chromosome 3P21.31 was under positive natural selection in East Asians. Molecular Biology and Evolution, 31(3), 683–695. https://doi.org/10.1093/molbev/mst260

Edwards, C. L., Fillingim, R. B., & Keefe, F. (2001). Race, ethnicity and pain. Pain, 94(2), 133–137. https://doi.org/10.1016/s0304-3959(01)00408-0

Fu, Q., Hajdinjak, M., Moldovan, O. T., Constantin, S., Mallick, S., Skoglund, P., Patterson, N., Rohland, N., Lazaridis, I., Nickel, B., Viola, B., Prüfer, K., Meyer, M., Kelso, J., Reich, D., & Pääbo, S. (2015). An early modern human from Romania with a recent Neanderthal ancestor. Nature, 524(7564), 216–219. https://doi.org/10.1038/nature14558

Hoffman, K. M., Trawalter, S., Axt, J. R., & Oliver, M. N. (2016). Racial bias in pain assessment and treatment recommendations, and false beliefs about biological differences between blacks and whites. Proceedings of the National Academy of Sciences, 113(16), 4296–4301. https://doi.org/10.1073/pnas.1516047113

Johansson, S. (2014). The thinking Neanderthals: What do we know about Neanderthal cognition? WIREs Cognitive Science, 5(6), 613–620. https://doi.org/10.1002/wcs.1317

Metzl, J. M., Petty, J. L., & Olowojoba, O. V. (2018). Using a structural competency framework to teach structural racism in pre-health education. Social Science & Medicine, 199, 189–201. https://doi.org/10.1016/j.socscimed.2017.06.029

Neary, M., & Owen, A. (2017). Pharmacogenetic considerations for HIV treatment in different ethnicities: An update. Expert Opinion on Drug Metabolism & Toxicology, 13(11), 1169–1181. https://doi.org/10.1080/17425255.2017.1391214

Neighbors, H. W., Caldwell, C., Williams, D. R., Nesse, R., Taylor, R. J., Bullard, K. M. K., Torres, M., & Jackson, J. S. (2007). Race, ethnicity, and the use of services for mental disorders. Archives of General Psychiatry, 64(4), 485. https://doi.org/10.1001/archpsyc.64.4.485

Race, Ethnicity, and Genetics Working Group. (2005). The use of racial, ethnic, and ancestral categories in human genetics research. The American Journal of Human Genetics, 77(4), 519–532. https://doi.org/10.1086/491747

Reardon, S. (2016). Neanderthal DNA affects ethnic differences in immune response. Nature. https://doi.org/10.1038/nature.2016.20854

Seeman, N. C. (2003). DNA in a material world. Nature, 421(6921), 427–431. https://doi.org/10.1038/nature01406

Wall, J. D., Yang, M. A., Jay, F., Kim, S. K., Durand, E. Y., Stevison, L. S., Gignoux, C., Woerner, A., Hammer, M. F., & Slatkin, M. (2013). Higher levels of Neanderthal ancestry in East Asians than in Europeans. Genetics, 194(1), 199–209. https://doi.org/10.1534/genetics.112.148213

Yamamoto, N., Yamamoto, R., Ariumi, Y., Mizokami, M., Shimotohno, K., & Yoshikura, H. (2021). Does genetic predisposition contribute to the exacerbation of COVID-19 symptoms in individuals with comorbidities and explain the huge mortality disparity between the east and the west? International Journal of Molecular Sciences, 22(9), 5000. https://doi.org/10.3390/ijms22095000

Özdemir, B. C., & Dotto, G.-P. (2017). Racial differences in cancer susceptibility and survival: More than the color of the skin? Trends in Cancer, 3(3), 181–197. https://doi.org/10.1016/j.trecan.2017.02.002

Chapter 10

Banks, W. E., d'Errico, F., Peterson, A. T., Kageyama, M., Sima, A., & Sánchez-Goñi, M. F. (2008). Neanderthal extinction by competitive exclusion. PLoS ONE, 3(12). https://doi.org/10.1371/journal.pone.0003972

Chiarelli, B. (2004). Spongiform encephalopathy, cannibalism and Neanderthals extinction. Human Evolution, 19(2). https://doi.org/10.1007/bf02437496

Degioanni, A., Bonenfant, C., Cabut, S., & Condemi, S. (2019). Living on the edge: Was demographic weakness the cause of Neanderthal demise? PLoS ONE, 14(5). https://doi.org/10.1371/journal.pone.0216742

Finlayson, C., & Carrión, J. S. (2007). Rapid ecological turnover and its impact on Neanderthal and other human populations. In Trends in Ecology and Evolution (Vol. 22, Issue 4). https://doi.org/10.1016/j.tree.2007.02.001

Gilligan, I. (2007). Neanderthal extinction and modern human behaviour: The role of climate change and clothing. World Archaeology, 39(4). https://doi.org/10.1080/00438240701680492

Gilpin, W., Feldman, M. W., & Aoki, K. (2016). An ecocultural model predicts Neanderthal extinction through competition with modern humans. Proceedings of the National Academy of Sciences of the United States of America, 113(8). https://doi.org/10.1073/pnas.1524861113

Goldfield, A. E., Booton, R., & Marston, J. M. (2018). Modeling the role of fire and cooking in the competitive exclusion of Neanderthals. Journal of Human Evolution, 124. https://doi.org/10.1016/j.jhevol.2018.07.006

Smith, F. H., Janković, I., & Karavanić, I. (2005). The assimilation model, modern human origins in Europe, and the extinction of Neandertals. Quaternary International, 137(1). https://doi.org/10.1016/j.quaint.2004.11.016

Timmermann, A. (2020). Quantifying the potential causes of Neanderthal extinction: Abrupt climate change versus competition and interbreeding. Quaternary Science Reviews, 238. https://doi.org/10.1016/j.quascirev.2020.106331

Trinkaus, E., & Howells, W. W. (1979). The Neanderthals. Scientific American, 241(6). https://doi.org/10.1038/scientificamerican1279-118

Underdown, S. (2008). A potential role for Transmissible Spongiform Encephalopathies in Neanderthal extinction. Medical Hypotheses, 71(1). https://doi.org/10.1016/j.mehy.2007.12.014

www.ingramcontent.com/pod-product-compliance
Lightning Source LLC
Chambersburg PA
CBHW022108160426
43198CB00008B/398